10 experiments in astronomical spectroscopy

for amateurs

Olivier Garde

edp sciences

This book is based on data produced by the author in various observatories: Observatoire de Haute-Provence, C2PU Calern (Observatoire de la côte d'Azur, Astroqueyras, Mars Observatory in Ardèche France, Observatoire de la Tourbière in Isère region in France, Observatoire 2SPOT in Chile hosted by Deep Sky Chile.

All illustrations and photos in this book have been produced by the author unless otherwise stated.

Cover picture: Horsehead nebula in the constellation of Orion, HII region called IC434. Acquisitions with a CCD SBIG STX 16803 (LRVB filters) and Ritchey-Chrétien telescope of 400mm in diameter.

Made in LaTex according to the template proposed by Mathias Legrand (Orange Book template) with modifications by Vel, under licence CC BY-NC-SA 3.0

ISBN (paper): 978-2-7598-4009-0 – ISBN (ebook): 978-2-7598-4010-6

Contents

About the author

Olivier Garde discovered astronomy at the age of 10, when his parents gave him a Newtonian 115/900 telescope for his birthday. Like many people, he was totally impressed by the eyepiece view of Saturn and its rings.

Olivier's visual introduction to astronomy soon led to imaging. He started with film photography and later moved to digital imaging when the first CCD cameras became available (Audine, SBIG ST6).

Olivier has also travelled the world to observe total and annular solar eclipses, taking photographs and videos which have been broadcast by various media (Venezuela, Chile, South Africa, Australia, Mongolia, Egypt, Zimbabwe, Spain, Iceland, USA).

In 1996, he obtained a University Degree in Astronomy and Astrophysics at the University of Paris Sud XI, and also became interested in photometry, taking light curves of asteroids and variable stars.

In 1998, he joined the Lyon Ampère Astronomy Club (CALA) and met many amateurs who taught him a great deal about how to observe through a telescope.

In 2000, Olivier discovered spectroscopy thanks to Valérie Desnoux and Christian Buil. It was a great moment for him to see what could be measured with this technique.

In May 2003, he was able to take part in a major and seminal meeting organized by the CNRS (French National Centre for Scientific Research) and, in particular, by Jean-Pierre Rozelot: " Les rencontres d'Oléron " where professional and amateur astronomers met to discuss pro/am collaborations. It was on this occasion that the first pro/am collaborative project in spectroscopy was born: Be-type stars and the creation of a dedicated spectral database: BeSS (Be Stars Spectra), set up by Coralie Neiner at the Paris-Meudon Observatory. For him, it was a major realization of what amateurs could contribute to science at their own level, a bit like the year zero of amateur spectroscopy, even if a few precursors had already started well before this meeting, but it did help to federate a few professionals and amateurs around a joint work project, particularly at the time around the Be stars.

He has also taken part in many astronomical observations on research telescopes, such as the Saint-Véran observatory (Astroqueyras - France), at the Mars observatory in Ardèche (France), at C2PU to the Côte d'Azur observatory on the Calern plateau (close to the french riviera), at the OHP (Observatoire de Haute-Provence) with the 1.52 m telescope.

Fig. — The 2003 Oléron pro/am meetings. Photo of the group of amateurs and professionals at these meetings organized by the CNRS in May 2003.

In 2017, during a workshop on Be stars at the Paris-Meudon observatory, he joined the team of validators of spectra of the BeSS database.

In 2018, Olivier joined the team at Shelyak Instruments (French manufacturer of spectrographs) as a technical consultant.

In 2019, Olivier and collaborators Stéphane Charbonnel, Pascal Le Dû, Lionel Mulato and Thomas Petit, founded 2SPOT (*Southern Spectroscopic Project Observatory Team*). 2SPOT operates 2 remotely piloted telescopes dedicated to spectroscopy in Chile, one in low resolution with an Alpy 600 spectrograph, the second with an Echelle (eShel) spectrograph in medium resolution. Extensive automated operation and processing software development has made 2SPOT an exceptionelly productive facility.

In 2023, together with Christian Buil, Valérie Desnoux, Matthieu Le Lain and Guillaume Bertrand, they set up the STAROS Projects association, with the aim of launching spectroscopic measurement campaigns and developing new tools and instruments. Thus, in november 2024, they created the SUNSCAN, an all-in-one spectroheliograph to scan the Sun at various wavelengths.

Olivier is also involved in several pro/am collaborations in spectroscopy, and has co-authored numerous publications.

Acknowledgements

My first thanks go to my parents Louis and Annie who got me interested in history, science, nature and music from an early age. Without them, I would probably never have taken up astronomy.

This book would never have seen the light of day if, in the course of my life, I had not met the people I am going to mention now, starting with my wife Florence, who has supported me now for over 20 years, but who, as a scientist herself, gave me a lot of pertinent feedback on how to write this book.

To my brother François, who is also my partner in an audiovisual company that we created and have been running together for over 30 years, where we produce many scientific videograms. Our collaboration is always in a spirit of innovation.

Since 1997, within CALA (Club d'Astronomie de Lyon Ampère), I have also met many people who have enabled me to take an interest in aspects of amateur astronomy that I knew little or nothing about. So I would like to thank the members of CALA, and especially:

Pierre Farissier, its president, Sophie Combe, its vice-president, Jean-Paul Roux, its secretary, and the members I have met over the years, such as Marcel Astier, Pascal Ayrault, Kévin Bazan, Laurent Béranger, Franck Bompaire, Bruno Christmann, François Cochard, Christophe Gillier, Serge Golovanow, Bertrand Guégan, Frédéric Hembert, Hélène et Christian Hennes, Vincent Lecocq, Jean-Pierre Masviel, Jacques Michelet, Régis et Angélique Nicolas, Céline Petit, François Régembal, Serge Renard, Christian Revol, Jean-Charles Ripault, David Romeuf, Éric Sibille, Mathilde Silvestre, Yvan Soubeyrat, Georges Termignon, Lucas Thibaud, Olivier Thizy, Kevin Tibi, Jean-Louis Tressol, Adrien Viciana, Andréas Zeinert, and many others (I apologize for those who are not mentioned, I am sure I am forgetting some).

Thanks to Pierre Dubreuil and Alain Lopez, who introduced me to the Calern site and the two 1m telescopes of the C2PU where I was able to observe with the LISA and Whoppshel spectrographs.

Many thanks also to Joe Daglen and Ken Hudson for organizing the big spectroscopy meeting in 2019 in Las Cruces, New Mexico (SMSW-2: Sacramento Mountains Spectroscopy Workshop). It was a great opportunity to meet many amateurs and professionals all around the world.

My thanks also go to several astrophysicists and astronomy professionals, such as

Agnès Acker, Denis Gillet, Jaroslav Merc, Coralie Neiner, Jean-Pierre Rivet, Vincent Robert, Alexandre Santerne, Steven N. Shore who have all given up their time to integrate amateurs into their professional scientific research projects, providing us with relevant information in certain fields of astrophysics or helping us promote pro/am actions in astronomy. Without them, amateur spectroscopy would not be as advanced, efficient or useful.

Shelyak Instruments, founded in 2006, has also played a major role in motivating me to take up spectroscopy with the equipment it has designed for amateurs (and professionals too). Many thanks to François Cochard and Olivier Thizy for creating this unique company.

Thanks also to Daniel Vérilhac, who introduced me to the RC 600 at Mars Observatory in Ardèche (France). Together we made a number of spectra on this telescope. Daniel also did a thorough re-reading of my book, a long and tedious job.

Thanks also to Christian Buil, Stéphane Charbonnel, Pascal Le Dû and François Teyssier for proofreading certain chapters. Their pertinent comments have been taken into account.

Thanks to Hamish Barker, Jonathan Dancer, Ken Harrison, Ken Hudson, Robin Leadbeater, Tom Love and Peter Velez, who kindly checked my English translation of the book and provided pertinent comments on certain paragraphs or chapters. Their contribution was indispensable to the writing of this English version.

I can not end these words of thanks without mentioning in particular Christian Buil and his wife Valérie Desnoux, both of whom have been involved in spectroscopy for many decades, the pioneers of amateur spectroscopy. They have greatly motivated me to take up this discipline, as have many other amateurs.

And then there is the whole team behind 2SPOT (Southern Spectroscopic Project Observatory Team): Stéphane Charbonnel, Pascal Le Dû, Lionel Mulato, Thomas Petit. Together, we have led an ambitious and exciting project to set up two telescopes and spectrographs in Chile controlled from a distance.

Then more recently, with Matthieu Le Lain and Guillaume Bertrand, two computer geniuses, together with Christian Buil and Valérie Desnoux, we created STAROS Projects for the management of spectroscopic databases and the development of tools and hardware for amateur spectroscopy.

1. Introduction

1.1 The goal of this book

The field of investigation in astronomical spectroscopy is vast. In this book, we aim to show just part of the diversity of measurements that can be made in this astronomical domain, whether measuring velocities, the evolution of a profile over time, or the spectral signature used to classify an object.

This book describes 10 practical projects that provide specific astronomical measurements. For each project we describe the equipment used, the method of acquisition and the free software used to complete the data processing.

In the following chapters, we will discuss:

1) Star temperature and spectral classification, determine whether it's a dwarf star or a supergiant.

2) The distance of galaxies and quasars, spectral shift measurement using the Doppler-Fizeau effect and measuring their escape velocity.

3) Spectro-imaging of planetary nebulae, how to digitize the surface of these objects and and take measurements with certain lines in the spectrum (electron density and temperature).

4) Studying Be-type stars, track their evolution over time and detect outburst.

5) Studying novae and supernovae by analyzing the evolution of their spectra, measure

the ejecta's expansion velocity. Determine the type of supernova based on spectral line present. presents.

6) How to confirm the nature of objects not yet referenced such as candidate planetary nebulae.

7) Measuring planetary rotation velocity, galaxies rotation velocity, disks of matter rotating around certain stars or accretion disks close to a black hole.

8) Creating time series over a short period of time and highlight spectrum changes on a particular line.

9) Studying spectroscopic binary stars, determine their periods, highlight the presence of one or more companions.

10) Measuring astronomical filters, slits and liquids using a spectrograph "on table" in the lab: the Lab'Ex.

Unless otherwise stated, all graphs, spectra, drawings, photographs and diagrams in this book were produced by the author.

1.2 Prerequisites in astronomy

To be able to carry out the various experiments proposed in this book, the reader must already have mastered:

• Setting up an equatorial mount by aligning it with the north or south celestial pole at less than 2 arc minutes. This precision is required to point at the right target and achieve precise autoguiding tracking.

• Precise pointing to a target to position it on the spectrograph slit.

• The acquisition software: for shooting with a CMOS or CCD camera, gain management for CMOS cameras. Note: it is advisable to acquire spectra in 1x1 binning with a CMOS camera.

• Doing autoguiding on a set point X, Y, on the target itself or on a neighboring star and visible on the autoguider's sensor.

• Setting the acquisition PC's clock to provide precise dating for each spectrum produced (to be carried out before each observation).

• Adjusting your spectrograph, during daylight and on a table to produce the best possible spectra (focusing and orientation of the spectral camera, focus and orientation of the autoguiding camera).

• Spectrum processing (calibration in wavelength and intensity). Certain specific processing procedures and some useful clarifications on certain points will be described in this book.

These prerequisites should pose no problems for those who are already involved in deep-sky imaging. For spectra processing, software can be found on the Internet and there are plenty of tutorials on how to do it.

1.3 Units, conventions and physical values in spectroscopy

In this book, you will discover a number of physical values, conventions and units. Here is a description of some of the most commonly used:

1.3.1 Wavelengths

The wavelengths are expressed in ångströms (Å). It is a “historical” tradition in astronomical spectroscopy to use this unit when it would have been more appropriate to use IS units (international system): the nanometer noted nm or the micrometer noted µm with:

$$1\ \mu\text{m} = 1\,000\ \text{nm} = 10\,000\ \text{Å}$$

By convention, short wavelengths are placed to the left (the blue part of the spectrum) and long wavelengths on the right right (the red part).

1.3.2 Spectral resolution of a spectrograph

In spectroscopy, we often indicate the spectral resolution denoted R. It corresponds to the ratio between the observed wavelength and the full width at half maximum (FWHM) of a spectral calibration line measured in the same area (neon, argon/neon or thorium/argon lamps are probably the most common types). The power of resolution R is given by the following equation:

$$R = \frac{\lambda}{\Delta\lambda}$$

with:
λ: the wavelength over which resolution is calculated.

$\Delta\lambda$: the full width half maximum of an atmospheric line or calibration reputed to be "very thin" (FWHM), This concept is described in detail in section 8.4.

The higher the R, the more resolving the spectrograph and the possibility to show more precise details: very slow radial velocities, exoplanets or small changes in the spectrum. For example, for a $\Delta\lambda$ (FWHM) of 1.5 Å and a wavelength observed λ of 6 563 Å, we obtain a resolution

$$R = \frac{6\,563}{1.5} = 4\,375.33$$

- For $600 < R < 3\,000$, we are in the low-resolution domain.
- For $10\,000 < R < 20\,000$, we are in the medium resolution domain.
- With $R > 20\,000$, we are in high-resolution.

The limits indicated are purely arbitrary and depending on the publication, the definition of high, medium or low resolution may differ. Some professional spectrographs offer R resolution over 70 000. In the case of amateur spectrographs, the resolutions R range from around 100 to 30 000.

1.3.3 Wavelength ranges

For the purposes of this work, the wavelength domain is divided into 3 distinct zones:

- The visible domain (from 400 to 750 nm) which corresponds to what the human eye can see.

- The near infrared, IR (from 750 nm to around 1 μm).

- The near ultraviolet, UV (from 340 to 400 nm).

These limitations are essentially due to the capabilities of the sensors available to amateurs. There are, of course, professional cameras capable of capturing wavelengths other than the visible domain, but at prohibitive cost to the amateur astronomer.

1.3.4 The *ADU*

The term "*ADU*", for *Analog to Digital Unit*, is often used to indicate the signal level present on a pixel of a CCD or CMOS camera. Each photosite converts the light received in electrons, into an analogue signal, then as a digital signal (*ADU*). The maximum level possible of *ADU* (ADU_{Max}) depends on the dynamic range N of the sensor (generally 8, 12, 14 or 16 bits) with the relation:

$$ADU_{Max} = 2^N$$

For example, a sensor with a 16-bit dynamic range may have 65 536 different values (2^{16}) which correspond to shades of grey ranging from 0 = black (no signal) to 65 536 = white (maximum signal value). If a signal is greater than the maximum possible *ADU*, it will peak at this value. In this case, the pixel becomes saturated. It is therefore advisable not to exceed 2/3 of maximum capacity, which keeps the signal within the linearity range and avoid the consequences of possible variations in sky transparency during an observation.

1.3.5 1D and 2D files

A 2D spectral file corresponds to the image of the spectrum as recorded by the spectral camera. It consists of a matrix of pixels on 2 axes: X and Y. Each pixel has an intensity whose range of values depends on the dynamic range of the sensor (see section 1.3.4).

A 1D spectral file is a spectral profile (curve) containing the wavelength value on the x-axis and the intensity of the spectrum on the y-axis.

These 2 files must be in .fit or .fits format (can also be in .dat format, but without the advantage of the fit header).

1.3.6 Binning area

The binning area of a spectrum corresponds to the height of the 2D spectrum on which pixels in the same column are added together. Each processing program allows you to define this area.

1.3.7 Sky background subtraction areas

In the same way as the binning area, two zones are defined outside the useful spectrum to subtract the value of the background sky from the spectrum to be studied. These zones usually contain the interstellar lines, atmospheric airglow and absorbtion lines and, as the case may be, light pollution lines. Spectrum processing software can be used to define these 2 areas.

1.3.8 Coordinates of a celestial object

The coordinates of an object in the sky are indicated in an equatorial reference frame by 2 values:

- The right ascension (α) noted RA in hours, minutes, seconds.

- The declination (δ) noted DEC in degrees, minutes, seconds.

These coordinates are usually followed by the term "J2000" which refers to coordinates calculated for the year 2000 (the positions of the stars move slowly across the sky!!!). JNow refers to coordinates at the time, take care to check witch software or telecospe controler is using.

1.3.9 Bias, darks, flats

As with astronomical imaging, calibration images are required:

Bias images are images taken with the camera shutter closed (complete black). They should be taken with the shortest possible exposure time: zero seconds. They are used to subtract read noise from the CCD or CMOS sensor matrix.

Dark images are also taken with the shutter closed. They must be made at the same temperature and have an exposure time greater or equal to the longest time exposure.

These images are used to subtract the thermal noise generated by the sensor depending on the exposure time, its temperature and the camera gain if the sensor is a CMOS.

Flat images: flats are images with uniform illumination, produced using a light source from a halogen lamp, with no spectral lines. As in imaging, Flats are used to eliminate the effects of dust on the sensor or vignetting in the optical chain.

In spectroscopy, they must be made in large numbers, in order to improve the signal-to-noise ratio because the flux of halogen flat lamp is very low in the blue part of the spectrum.

1.3.10 The F/D ratio

For a telescope or refractor, we often indicate the aperture ratio F/D, which corresponds to the ratio of its focal length to its diameter. For example, a telescope with a focal length of 1 500 mm and a diameter of 300 mm will have an F/D ratio of 5.

It is important to respect this aperture ratio, which depends on the spectrograph used, optimized to operate at a defined F/D ratio. However, it is possible to use a Barlow or focal reducer lens to adjust the F/D of the telescope or refractor used. For example, an Alpy 600 will perform optimally when used with a telescope or refractor open to F/5 or even F/4.

1.4 The spectrographs

There are many spectrographs within the financial reach of amateurs. This chapter describes the main models used by amateurs, but also by some professionals at the time this book was written. This list is not exhaustive.

As with telescopes, there is no such thing as a universal, all-purpose spectrograph. Depending on the field of study and the nature of the target, you need to be able to choose the one that's right for you. This is why we find a large range of instruments that differ greatly in terms of resolution or spectral range, fiber optic, with slit or slitless.

For example, we are not going to use the same instrument to study spectroscopic binaries or the redshift of galaxies.

By contrast, the study of spectroscopic binaries stars requires high spectral resolution. On the other hand, the accessible magnitude will limit us to the brightest stars (because the light is dispersed over a larger area).

Studying the redshift of quasars or galaxies requires the ability to take spectra of very faint objects while offering a wide range of wavelengths. In this case, a low-resolution spectrograph is the best choice.

There are several solutions available to those wishing to venture into astronomical spectroscopy.

1.4.1 Building your own spectrograph

You can build your own instrument using the plans that are freely available on the Internet. In recent years, the advent of 3D printers has made it possible to produce high-performance spectrographs at low cost.

In 2024, there are four 3D-printed spectrograph projects. Printable files can be found on the websites of their authors:

• The **UVEX 4**: a spectrograph designed by Christian Buil, which was taken over by the "Nice People Team" (Pierre Dubreuil, Alain Lopez, Jean-Luc Martin and Stéphane Ubaud) for the 3D-printed mechanical part. A comprehensive website is dedicated to this project: http://spectro-uvex.tech

• **Low Spec**: a spectrograph designed by Paul Gerlach. The project and the files are presented on the Thingiverse website: https://www.thingiverse.com/thing:2455390

The Sol'Ex and Star'Ex projects: designed by Christian Buil, which, depending on the configuration used, produce high-resolution spectra at $R = 20\ 000$ or low resolution at $R = 800$.

• The **Sol'Ex** is dedicated to spectroheliography (scanning the solar disk at various wavelengths).

• The **Star'Ex** allows you to make stellar spectra thanks to the addition of a slit and an autoguiding module.

A website dedicated to the construction and use of these 2 spectrographs can be consulted on Christian Buil's website: https://solex.astrosurf.com/

The Sol'Ex and Star'Ex optical kits are sold by Shelyak Instruments on their website.

• **FlexSpec**: an open source spectrograph project whose plans, codes and documentation can be found on the website: https://flexspec1.readthedocs.io/en/latest/

1.4.2 Buying a spectrograph

A number of instrument manufacturers are also on the market which offer a variety of ready-to-use models with different resolutions and spectral ranges:

• SBIG, a US manufacturer, launched the first commercial spectrograph, the SGS in the 2000s followed by the DSS-7. These 2 models are no longer in the manufacturer's catalog. The SGS was a particularly interesting model for users of SBIG ST7, ST8, ST9 and ST10 CCD cameras, because it used the 2 sensors of these cameras: one for autoguiding on the spectrograph slit, the second for spectrum acquisition. To this day, the SGS remains a high-performance spectrograph thanks to its all-mirror optical design. It offers 2 gratings (600 and 150 lines/mm) and you can switch from one to the other very easily. The SGS has 2 different slit widths (25 μm and 100 μm).

• Baader Planetarium, a German manufacturer, which markets the Dados and the BACHES Echelle spectrograph. Dados can be used with various gratings of 200, 900 and 1 200 lines/mm and various slit widths: 25 μm, 35 μm and 50 μm with a possible spectral range from 350 nm to 950 nm. BACHES is a Echelle-type, so a multi-order spectrograph. It provides the entire spectral range between 390 and 800 nm at high resolution ($R = 18\,000$).

Internet website: https://www.baader-planetarium.com/en/spectroscopy.html

• Starlight Xpress a UK manufacturer: they proposed the SX spectrograph in 2016, followed by the Spectrograph Pro. The grating used is 550 lines/mm. It has several slits of different widths to best adapt the resolution depending on telescope focal length. The resolution can vary from $R = 1\,500$ to $R = 1\,800$ according to the slit used. The possible spectral range is from 340 to 900 nm.

Internet website: https://www.sxccd.com/product/sx-spectrograph-pro/

• Shelyak Instruments, a French manufacturer, offers a wide range of high and low resolution spectrographs:

- Alpy 600 and Alpy 200
- LISA
- LHIRES III
- UVEX
- eShel

Internet website: www.shelyak.com

1.4.3 The spectrographs used in this book

Here are the characteristics of some commercial instruments used to produce the results presented in this book. Other spectrographs could have been used too with the same quality of results, but some of the spectra published in this book are more than 10 years old that could not be produced by more recent spectrographs.

They were made using the range available at this time. Many spectra have been acquired from our 2SPOT setup in Chile with the eShel and Alpy 600 spectrographs.

They are also the most widespread commercial versions in the world at the time of writing this book (2024).

1.4.4 The Star Analyser

Fig. 1.4.1 — Star Analyser SA 100.

This spectrograph is the simplest and least expensive on the market. It has the external appearance of a 31.75 mm diameter astronomical filter (1.25 inch) which can be screwed directly in front of the camera or place in a filter wheel.

It features a diffraction grating with a transmission grating of 100 or 200 lines/mm (SA 100 or SA 200). The Star Analyser disperses the flux of the light passing through the filter and produces a spectrum.

It simply mounts in front of the camera with extension rings to adapt spectrum dispersion on the camera sensor.

The SA 100 has a nominal resolution *R* of 100 and its big brother, the SA 200 has a resolution of 200. All you need to do is orient it correctly to obtain horizontal spectra.

It is ideal for amateurs who want to get started with spectroscopy at low cost and can be used to observe very faint targets, such as supernovae or quasars.

On the other hand, its low resolution and lack of a slit will quickly limit its potential use in astronomical spectroscopy.

Spectral processing remains a delicate issue especially if the area under study contains many stars on large targets (e.g. galaxies, planetary nebulae) or in a highly light-polluted environment.

1.4.5 The Alpy 600

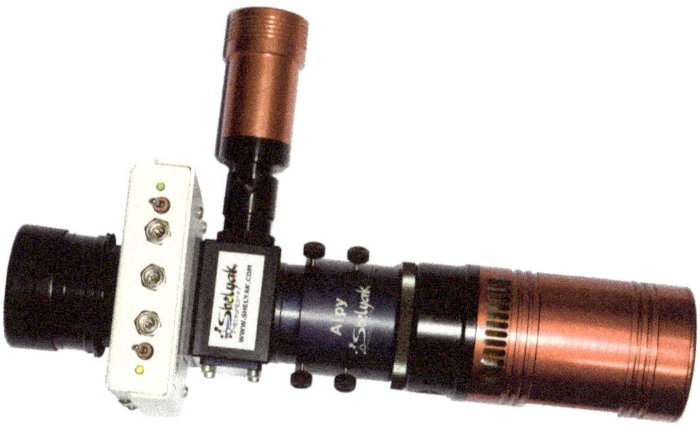

Fig. 1.4.2 — The Alpy 600. Equipped with its 2 cameras (spectral and autoguiding) as well as the calibration and autoguiding module.

It is with this spectrograph that we are starting to make some interesting measurements. As the Alpy 600 is modular, you do not have to buy all 3 modules up front (spectrograph + autoguiding module + calibration module).

We can start with the spectrograph body alone which makes it possible to study the sunlight spectrum, the spectrum of various light sources or measure the bandwidths of astronomical filters. It is equipped with a grism, a transmission grating glued to a prism which returns the light beam perpendicular to the sensor.

If you want to get started in celestial observation, then you need the autoguiding module which reflects the image of the slit onto a guide camera sensor to position the target correctly and hold it in place on the slit.

The calibration module completes the package for easy flat production and spectra from a calibration lamp (integrated halogen and argon/neon bulbs).

This spectrograph, equipped with a 23 μm slit, has a nominal resolution of $R = 600$, it is very compact and fairly lightweight. It can therefore be used on a small telescope or refractor without adding significant weight to the entire chain. Its compactness makes it less sensitive to mechanical bending and to temperature variations. This stability makes it a major asset for remote use. This is why it is used by our 2SPOT team in Chile.

This is the ideal low-resolution spectrograph for reflecting telescopes or refractors from 72 to 305 mm in diameter. On larger instruments, it loses some of its potential because we'll have to increase the slit width by 23 μm to 35 μm or even 50 μm,which reduces the resolution.

Its compact size means it can be used in "Lab'Ex" mode on table, to produce spectral measurements in laboratory mode on various light sources, filters, liquids or minerals. This possibility is described in Chapter 11 of this book.

A very low-resolution version (Alpy 200) is also available for very weak targets with a resolution of the order of $R = 100$ which can be used on supernovae, for example. Externally, it is totally similar to an Alpy 600, only the grism changes.

1.4.6 The LISA

The LISA spectrograph has a nominal resolution of $R = 1\,000$ thanks to its high luminous efficiency (it acts as an internal focal reducer). It is dedicated to the study of weak objects like supernovae, novae, candidate planetary nebulae.

It has its own calibration module with a halogen lamp for flats and an argon/neon or neon-only lamp for spectral calibration. Designed for use in the visible range, its spectral range extends from 390 nm to 750 nm. A near-infrared version is also available, providing spectra down to 1 μm.

A autoguiding module is integrated into the spectrograph eand any C-mount or C/S-mount camera can be attached (back focus of 12.5 mm or 17.5 mm).

It is suitable for use with telescopes of 305 mm in diameter and larger. LISA currently operates on telescopes from 0.6 m to 1 m in diameter, by producing spectra of high-magnitude objects. We used it on the C2PU 1m telescope at the Côte d'Azur observatory (close to the french riviera) and on the 0.6 m telescope at the Mars Observatory in Ardèche in France.

Of course, the width of the slit will have to be adapted according to telescope focal length. The larger it is, the wider the slit needs to be. Other slits are available as options, ranging from 10 μm to 100 μm. The trade off is a reduction in spectral resolution which will be around $R = 700$ with a 35 μm slit or $R = 500$ with a 50 μm slit.

Fig. 1.4.3 — The LISA. Equipped with 2 cameras (spectral and autoguiding) and the calibration module (in white on the photo). In this picture, it is also equipped with a radio module for remote control of calibrations lamps.

1.4.7 The UVEX

In addition to the 3D-printed version of UVEX 4 mentioned in section 1.4.1, an industrial version is available from Shelyak Instruments. It is a pretty versatile instrument which offers an extended wavelength range (from near UV around 340 nm to 1 μm in the near IR).

This is made possible by the use of mirrors rather than lenses, so there's no chromatism in this spectrograph, which allows it to go quite far into extremes, without optical distortion.

It can be used at different resolutions by simply changing the the grating (150, 300, 600, 1 200 ou 1 800 lines/mm), to obtain resolutions around between $R = 300$ to $R = 6000$ on the $H\alpha$ line. A wide range of slits is also available.

A motorized version is available as an option to remotely control the spectrograph's internal focus, grating orientation and the 2 calibration lamps. It uses the Alpy 600 modules for autoguiding and calibration. The spectrograph can therefore be operated perfectly by remote control, either via the ASCOM driver designed for this motorization, or via a dedicated application.

Because of its optical design (Czerny-Turner), it is hardly possible to produce spectra of non-point objects because the astigmatism of this spectrograph is quite high as you move away from the optical axis. On the other hand, it will express its full potential on pointlike stellar targets even at high magnitudes.

Fig. 1.4.4 — The UVEX. Alone, without motorization, calibration and autoguiding module.

1.4.8 The LHIRES III

One of the world's oldest high resolution spectrographs! It was conceived in 2003 during CNRS meetings between professionals and amateurs in Oléron (on the French west coast).

This spectrograph is equipped as standard with a holographic grating of 2 400 lines/mm and achieves a resolution of $R = 17\,000$ with a 23 µm slit.

A micrometer screw adjusts its inclination and thus vary the observable wavelength range in the visible range. Various gratings can be easily mounted on this spectrograph and enable lower resolutions (gratings of 1 200, 600, 300 ou 150 lines/mm).

Fig. 1.4.5 — The LHIRES III. Equipped with 2 cameras (spectral and autoguider).

Fig. 1.4.6 — The eShel. Equipped with its spectral camera (an ATIK 460EX).

It has been optimized for use with F/10 Schmidt-Cassegrain telescopes. It is equipped with an argon/neon calibration lamp and a halogen lamp for flats production. Unlike the Alpy 600, everything is integrated in this spectrograph (autoguiding and spectral calibration).

For autoguiding, any C-mount or C/S-mount camera can be attached (a 12.5 mm or 17.5 mm back focus).

With the Star'Ex HR, this is the spectrograph that delivers the highest spectral resolution in the entire range described in this chapter. Many amateurs own one, some have modified it to be motorized (wavelength range and internal focus of lens doublet).

1.4.9 The eShel

A Echelle-type spectrograph, i.e. it does not just produce a single linear spectrum but a series of orders that cover the entire visible range with an average resolution of about $R = 11\ 000$.

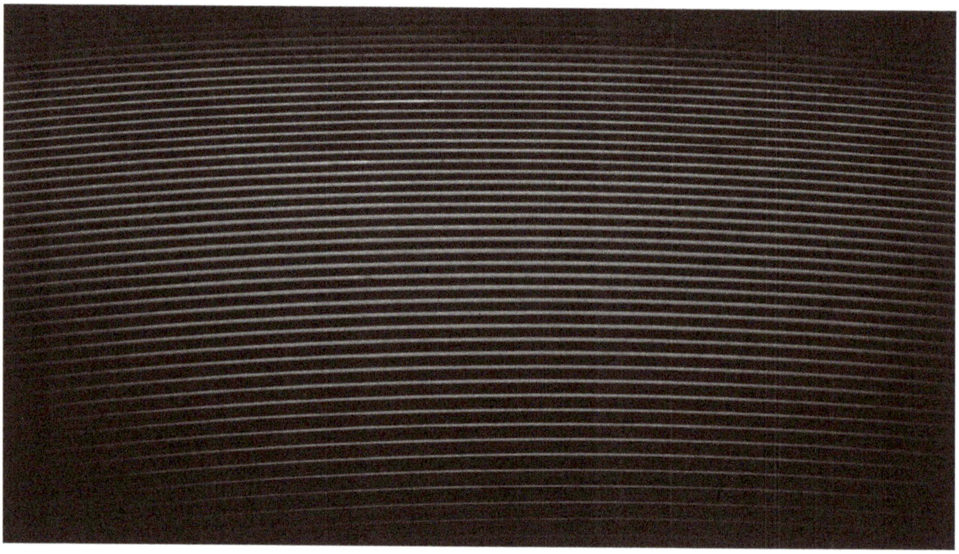

Fig. 1.4.7 — 2D Echelle spectrum. Raw spectrum of Beta Lyra using an eShel spectrograph $R = 11\ 000$. Each line corresponds to an order, the entire visible range is covered in a single exposure.

This spectrograph is not attached directly to the telescope but is connected to it by an optical fibre. The instrument is therefore not subject to mechanical flexion, and even better: it can be installed in a temperature-controlled enclosure for improved measurement stability.

1.5 Spectrography software

Here is a list of the most popular free spectra processing software. There are many other programs available, but some of them are difficult to use, others do not offer spectral analysis functions, only the simple processing of a spectrum. Not all have been tested, but here are the most common and widely used by amateurs.

1.5.1 Visual Spec

Software designed by Valérie Desnoux (a French amateur astronomer), it is very comprehensive, with advanced spectral analysis tools (mathematical operations, Gaussian function, spectral line identification, etc.) as well as the ability to check its spectral calibration and accuracy. It is very well documented. The software is available at the following link:

http://www.astrosurf.com/vdesnoux/

1.5.2 ISIS

Software designed by Christian Buil. It is versatile and very powerful. It can process spectra from the vast majority of common instruments including Echelle-type multi-order spectra. Many tools are available to perform calculations between different spectra, measure the width of a line, create spectra animations and many other functions. It can be downloaded from the following link:

http://www.astrosurf.com/buil/isis-software.html

1.5.3 Demetra

Designed by Nicolas Durand and François Cochard at Shelyak Instruments, it is the only software that lets you acquire spectra as well as process them. Dedicated to the range of spectrographs marketed by Shelyak Instruments, it processes Alpy 600, UVEX, LISA and eShel spectra. It also controls the various calibration modules for these spectrographs.

However, it has no spectral analysis tools nor the ability to perform arithmetic operations with spectra. Very easy to use, it is ideal for beginners. Download it here:

https://www.shelyak.com/logiciel/logiciel-demetra/

1.5.4 Spec INTI

Designed by Christian Buil and Valérie Desnoux (for the spec INTI editor part), it is the latest software release in 2019, it uses a Python environment. It has been specially developed for the Star'Ex spectrograph, but can also process single-order spectra from other instruments. It is available on this link:

http://www.astrosurf.com/solex/specinti1_fr.html

2. The temperature of stars

In this chapter, we will discover the different spectral classes by observing various stars whose magnitudes range from zero to 4. This is a very simple exercise to perform with a low-resolution spectrograph such as the Star Analyser, Alpy 600, Star'Ex or LISA. You can also use a LHIRES III or a UVEX, installing a grating of 150 to 300 lines/mm, which gives you the entire spectrum in the visible range between 400 nm and 750 nm, or even more depending on the spectral resolution and the camera used. As we will produced spectra of bright stars, long exposure times are not required.

2.1 The spectral classes

The first attempts at classification date back to the 19th Century when astronomers tried to classify stellar spectra according to dominant colours and certain visible lines. It was an Italian astronomer, Angelo Secchi (1818-1878), who first proposed a system of spectral classification in 1866. He created 3 classes:

• Class I: composed of blue or white stars with broad, pronounced absorption lines (the hydrogen lines). Examples are Vega, Altair, Deneb, Rigel, Bellatrix, Sirius, Algol, Atlas and Fomalhaut.

• Class II: made up of yellow stars with many faint lines. This is the case, for example, with the Sun, Capella, Arcturus, Arneb, Alshain and 16 Cygni B.

• Class III: composed of red stars with a multitude of lines that can be found on Antares, Betelgeuse and Proxima Centauri.

Angelo Secchi would later add, in 1868, a class IV of carbonaceous stars with obvious line patterns, then in 1877 class V stars with one or more emission lines such as gamma Cas, delta Sco, beta Lyr and Pleione (the future Be star class).

At the beginning of the 20th century, at Harvard University, Henry Draper (1837-1882) began compiling a catalog of more than 220 000 stars up to magnitude 9. Subsequently Edward Charles Pickering (1846-1919) came up with the idea of classifying stars according to their temperature. This tedious work was largely carried out by Williamina Fleming (1857-1911), a woman working at Harvard University. Initially, she was recruited by Edward Charles Pickering as a housekeeper, then entrusted by him with the thankless task of studying photographic plates to identify the various spectral lines. This work was later continued by 2 other women: Annie Jump Cannon (1863-1941) and Antonia Maury (1866-1952). The HD Catalogue was finally completed in the 1920s.

This temperature classification is still used today. Each class is named after a capital letter of the alphabet. Initially, it had been drawn up by type according to the lines that could be seen, but it did not follow a logical order and depended on several physical criteria at once.

To simplify this classification, it was decided to use just one criterion: the temperature of the stars. The result is a classification that no longer takes account of the original alphabetical order, and some letters have disappeared because they have been grouped together in the same spectral class. The 7 main classes are, in descending order of temperature:

O, B, A, F, G, K, M

A simple mnemonic to remember it:

Oh, **B**e **A** **F**ine **G**irl/**G**uy, **K**iss **M**e.

(there are many other phrases for memorizing this order of letters). You can invent your own too!

Here are the characteristic elements of each class: O is the highest temperature class (above 25 000 K) and M the coldest class (below 3 500 K):

O: T = 25 000 to 50 000 K. Very few visible lines apart from those of ionized helium.
B: T = 10 000 to 25 000 K. The spectrum shows helium and hydrogen lines.
A: T = 7 500 to 10 000 K. The spectrum shows mainly hydrogen lines (Balmer lines).
F: T = 6 000 to 7 500 K. Ionized metal lines such as calcium or iron are present along with Balmer lines.
G: T = 5 000 to 6 000 K. The number of metallic lines is higher than in previous classes.
K: T = 3 500 to 5 000 K. The spectrum is dominated by metallic lines.
M: T = 2 500 to 3 500 K. Strong molecular bands appear such as titanium oxide (TiO).

Each spectral class is divided into 10 subclasses denoted by a number from 0 to 9 to reflect the temperature value: 0 being the hottest subclass and 9 the coldest.

Finally, to complete the characterization of a star according to its spectral type, its absolute brightness (luminosity) which is proportional to its volume and density, must be taken into account. The luminosity classes are indicated after the spectral class by a Roman numeral from I to V. They can be found in the Herztsprung-Russell diagram (abbreviated to the HR diagram).

Ia: a-type supergiant stars.
Ib: b-type supergiant stars.
II: bright giant stars.
III: giant stars.
IV: sub giants stars.
V: small stars (dwarf stars) in the main sequence and denser than giants.

By examining the width of the hydrogen lines in a spectral profile, we can determine the size of the star.

Supergiants or giants rotate slower than dwarf stars. This causes the lines to be wider for a dwarf star.

On the other hand, supergiant or giant stars are less dense than dwarf stars. There are fewer atomic interactions than in a dwarf star, which affects the width of the lines: the more interactions there are between atoms, the wider the lines.

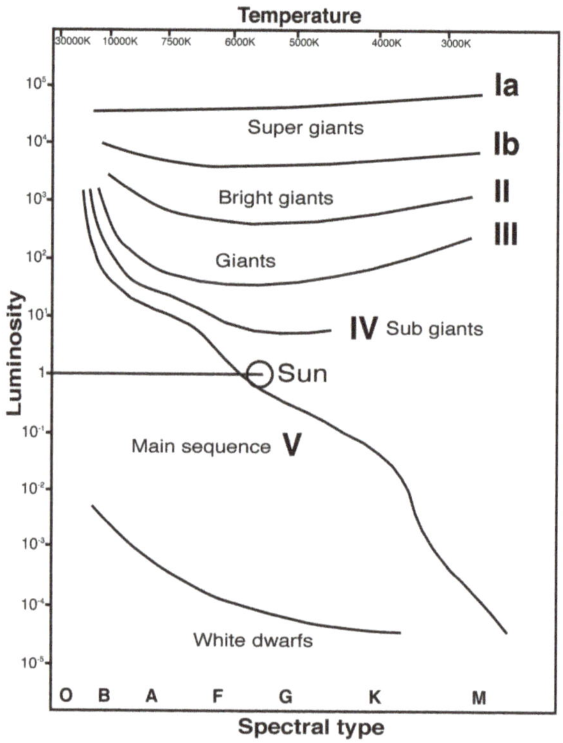

Fig. 2.1.1 — HR diagram. Herztsprung-Russell diagram showing luminosity classes as a function of temperature.

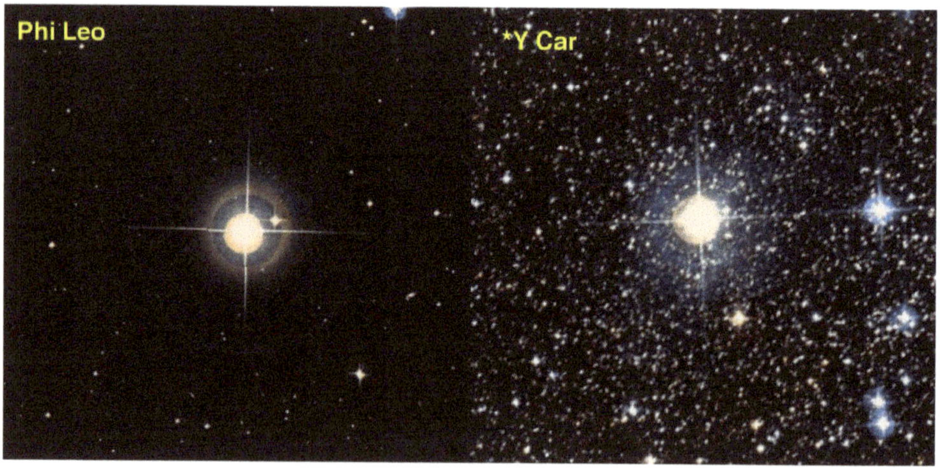

Fig. 2.1.2 — Comparison of luminosity classes Ia and V. Photography of phi Leo (A5V) and *Y Car (A5Ia). Picture credit: SIMBAD DSS-2 database.

Let's take the example of 2 stars of the same spectral class A5:

- phi Leo (HD98058) a dwarf star (V) of magnitude $V = 4.45$.
- *Y Car (HD97534) a supergiant star (Ia) of magnitude $V = 4.62$.

If we take a photograph of each of these 2 stars of very close magnitude, you get similar images in terms of colors (Figure 2.1.2), but if we obtain their spectra, we see 2 very different profiles (Figure 2.1.3).

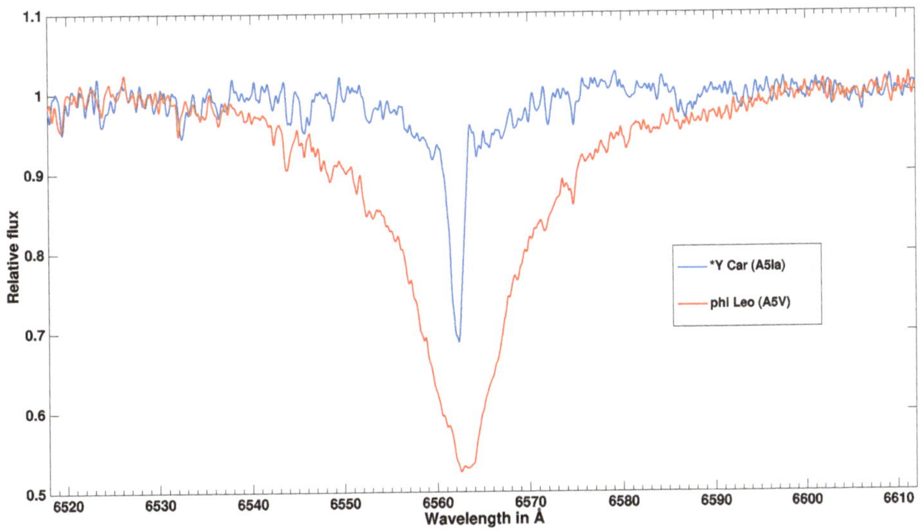

Fig. 2.1.3 — Comparison of luminosity classes Ia and V. Spectra of phi Leo (A5V) and *Y Car (A5Ia) centred on the $H\alpha$ line. The line of phi Leo (dwarf star) is much wider than that of *Y Car (Ia-type supergiant star) although these 2 stars belong to the same A5 spectral class.

The width of the line indicates whether it is a supergiant (luminosity class Ia or Ib) or a dwarf star (luminosity class V).

These types of profiles can be verified on other A or B type stars, which are easier to see than for cooler stars.

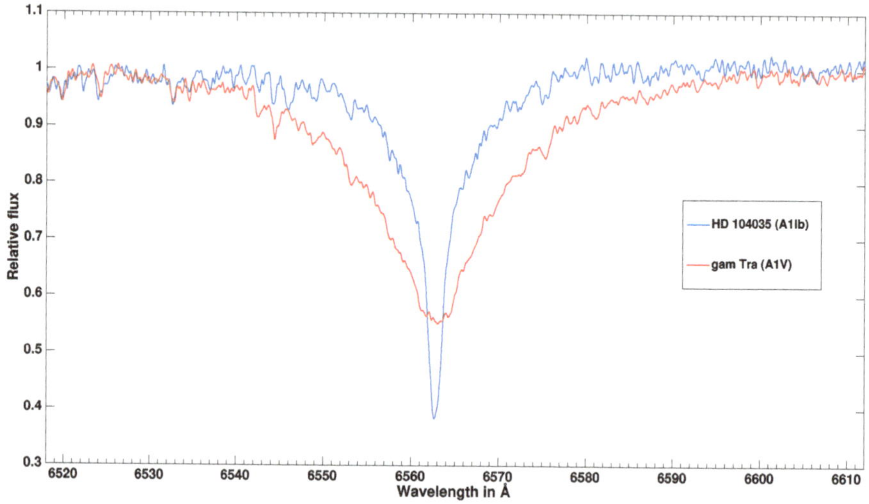

Fig. 2.1.4 — Comparison of luminosity classes Ib and V. Spectra of gam Tra (A1V) and HD 104035 (A1 Ib) centred on the $H\alpha$ line. The difference in profile width can be seen in the $H\alpha$ line, reflecting the luminosity class.

We saw in this chapter that there are 7 main types spectral (O, B, A, F, G, K, M). Subsequently other spectral types have been added, for example:

- C-carbon stars (very cold stars).

- Wolf-Rayet WR-type stars (whose gaseous envelope can be extremely hot, much hotter than an O-type star).

- The Be stars (which show one or more hydrogen lines in emission, reflecting the presence of a disk of material rotating around the star).

Other rare spectral types are also used, such as the L, T, Y, R, N and S classes, but these are not discussed in detail in this book.

2.2 Selecting targets

In order to produce spectra of stars of different types, the targets must be selected according to several criteria:

- Their magnitude: find a good compromise between exposure time and magnitude depending on the spectrograph used and the optics.

• Their visibility for a given date and site, well positioned in the sky (at least 30 degrees altitude).

• Their spectral class: the aim of the exercise is to produce a spectrum for each main type and a few others for special classes such as carbon stars or Wolf-Rayet stars.

To do this, we are going to use SIMBAD, an online tool for extracting information from Strasbourg Observatory's astronomical data center in France: the CDS.

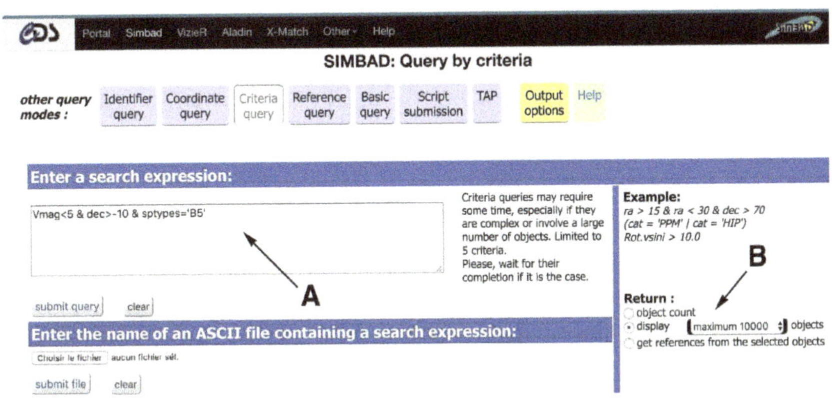

Fig. 2.2.1 — SIMBAD's query page.

1) Go to:

https://simbad.u-strasbg.fr/simbad/

2) Click on "**Criteria query**".

You get the page shown in Figure 2.2.1 where you can make precise requests. Please note! You can only request a maximum of 5 criteria.

If you want to find B5 type stars whose magnitude is less than 12 and whose declination is greater than -20 degrees, you will need to write in box A of the Figure 2.2.1, "**enter a search expression**", the following request:

Vmag<12 & dec>-20 & sptypes = "B5"

In the right column (zone B), start by checking "**object count**" which returns the number of objects matching the defined criteria.

This number can be very large. In the example shown in Figure 2.2.1, over 1 900 items are listed.

You will then need to refine your queries to reduce it; for example by requesting a lower magnitude, the previous request can become:

Vmag<5 & dec>-10 & sptypes = "B5"

When the number becomes reasonable, you can request the list of selected objects by ticking **"display"** instead of "**object count**". The list will be displayed, here the query returns 22 targets, as shown in Figure 2.2.2.
The operation will have to be repeated to choose your targets in other spectral classes by modifying the value **sptypes**. You can export this list in the Votable format or Ascii or CDS portal by selecting the option at the bottom of the results page.

C.D.S. - SIMBAD4 rel 1.8 - 2024.08.12CEST09:13:40
Vmag< 5 & dec >-10& sptypes='B5'
Number of objects : 22

#	identifier	type	coord1 (ICRS,J2000/2000)	Mag V	spec. type
1	* omi Cas	Be*	00 44 43.5186007265 +48 17 03.731037403	4.349490	B5IIIe
2	* phi And	Be*	01 09 30.12562 +47 14 30.4594	4,25	B5IIIe
3	* 57 Cyg	SB*	20 53 14.7587928897 +44 23 14.147866955	4.728380	B5V
4	HD 21278	SB*	03 28 03.0707602272 +49 03 46.325803896	4.954242	B5V
5	* psi Per	Be*	03 36 29.3797327450 +48 11 33.484405927	4.123160	B5Ve
6	* del Per	Be*	03 42 55.4928046977 +47 47 14.988257477	2.982313	B5III
7	* ome Cas	SB*	01 56 00.0283006680 +68 41 06.863913852	4.938846	B5III
8	* mu. Eri	SB*	04 45 30.1514281206 -03 15 16.775320589	4.008290	B5IV
9	* tau Ori	*	05 17 36.3903358088 -06 50 39.866785352	3.588380	B5III
10	* 32 Ori A	*	05 30 47.0510459520 +05 56 53.323013148	4.442832	B5IV
11	* 32 Ori	**	05 30 47.05091 +05 56 53.2925	4,2	B5IV+B7V
12	* f01 Ori	Be*	06 12 03.2799360825 +16 07 49.453623814	4.911102	B5Vn
13	* psi02 Aqr	Be*	23 17 54.2137206 -09 10 57.067524	4.402793	B5V
14	* omi Aqr	Be*	22 03 18.8440617830 -02 09 19.313835540	4.656648	B5V
15	* pi. And	SB*	00 36 52.8513216207 +33 43 09.636294654	4.304680	B5V
16	* nu. And	SB*	00 49 48.8463243776 +41 04 44.079221948	4.490910	B5V
17	* tau And	V*	01 40 34.8161856549 +40 34 37.339701270	4.926168	B5III
18	* iot Aql	*	19 36 43.2777028800 -01 17 11.758614396	4.306137	B5III
19	* 67 Oph	s*b	18 00 38.7157893000 +02 55 53.633402904	3.932415	B5Ib
20	* 22 Cyg	SB*	19 55 51.7609131816 +38 29 12.188844204	4.910045	B5IV
21	* lam Cyg	**	20 47 24.53766 +36 29 26.5737	4,54	B5V
22	* tau Her	Pu*	16 19 44.4367980146 +46 18 48.102135319	3.887130	B5IV

Fig. 2.2.2 — SIMBAD query results. List of 22 B5 spectral type stars.

2.3 Spectra production

Now that you have your target list, it is time to get hands-on with your astronomical equipment and produce the various spectra required. For this, the procedure to be used will always be the same and should follow the same series of steps:

1) Making FLATs at the beginning or end of the night. For a good signal-to-noise ratio

in the blue part of the spectrum, it is recommended to make at least thirty.
Note: adapted darks and bias, which vary little, and held in a library and can be reused.

2) Point at the 1st target and capture its spectrum. The unit exposure time and total number of exposures should be adjusted according to the magnitude of the target.

3) Obtain 2-3 exposures with a spectral lamp (to calibrate the wavelength spectrum of the target).

4) Point to a nearby reference star, at the same altitude as the target, to obtain an equivalent air mass and with the lowest possible color excess E(B-V) (calibration in intensity). To do this, we can use the spreadsheet program "Reference Star Finder" created by François Teyssier which is available on his website:

www.astronomie-amateur.fr

You can also search for an A or B type star using a map... or SIMBAD!

5) Start a few acquisitions of the reference star (adjust exposure time to obtain around 40 000 ADU).

6) Take 2-3 exposures with a spectral lamp (to calibrate the reference star).

Repeat all steps for all other targets. If you have chosen objects that are bright enough and well located in the sky, you should be able to complete all the observations in a single night.

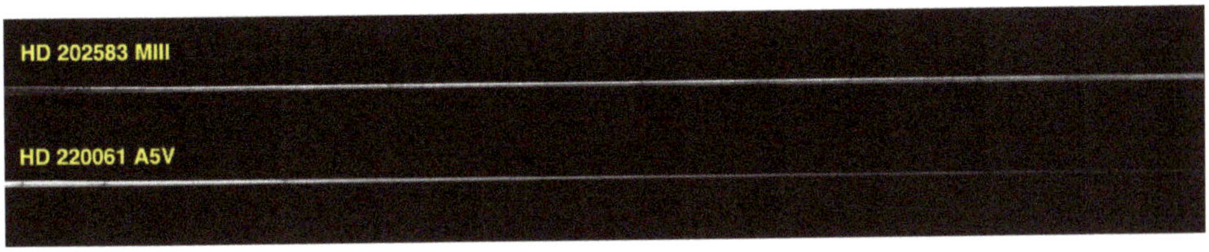

Fig. 2.3.1 — Example of two 2D spectra. Here are 2 2D spectra of above, a cold star: HD 202583, a MIII type star and below, a hot star: HD 220061, a A5V type star. These 2 spectra were produced using a LISA spectrograph: 7 x 60 second exposures on HD 202583 and 9 x 30 second exposures on HD 220061.

Your spectra can be processed with free software such as ISIS, Demetra, Visual Spec, Spec Inti and many others (see section 1.5 for a non-exhaustive list of software). The procedure for processing spectra is not explained here, because, on the one hand, there are numerous tutorials on the internet for each software package and on the other hand, the process is very different from one program to another. The choice of program depends on the type of spectrograph you are using but also your understanding of its capabilities or the ease with which you use it.

2.4 Highlighting the results

Once all your spectra have been processed, you will end up with several .fit 1D files. In this chapter, we will show you how to present them graphically and scientifically.

The graph of each spectrum can already be displayed using the functions available in the software. Most also offer the option of creating a .png graphic file (ISIS and Vspec use the free Gnuplot program to generate them).

You can also use Python or Matlab scripts (or its free equivalent: SciLab).
Matlab has dedicated functions for reading .fit or .fits files and their headers. There are also many Python libraries such as astropy to manipulate spectra .fit files and display them graphically.

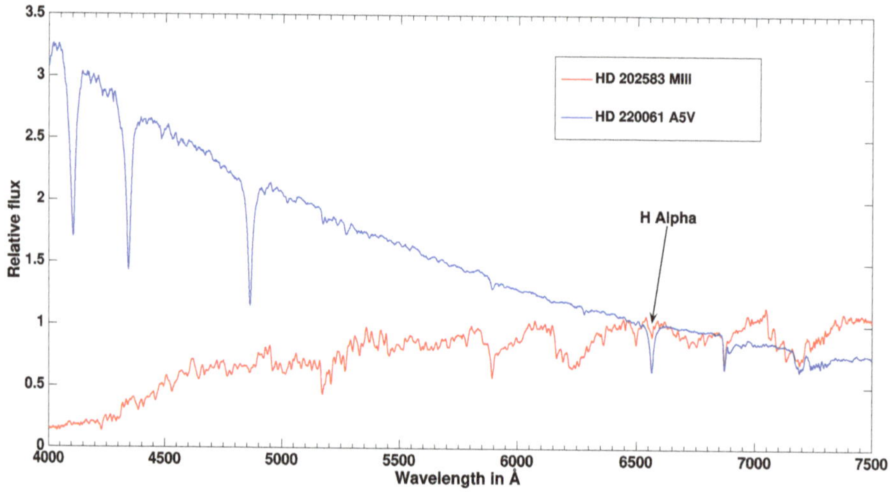

Fig. 2.4.1 — Graph of two 1D spectra. Spectra of a hot star HD 22006, A5V type (in blue) and of a cold star HD 202583, MIII type (in red). The two curves have been normalized to 1 close to the $H\alpha$ line area. Spectra taken with a LISA spectrograph (R = 700-800) mounted on a C14 SCT telescope equipped with an ATIK 414ex CCD camera.

By looking at the 2 spectra in Figure 2.4.1, we can see that the hottest star has its continuum rising towards the blue, while the coolest star has its continuum falling towards the blue and rising slightly towards the red.

To compare 2 spectra together, they should be standardized on the same wavelength range. Normalization consists of obtaining the arbitrary value of 1 in intensity on a portion of the continuum in a defined wavelength range with no lines. Software such as ISIS, spec INTI or Visual spec make this easy.

For example, in the ISIS software, display your spectral profile in the "**5-Profile**" tab, then select on the right the case "**Normalize**".

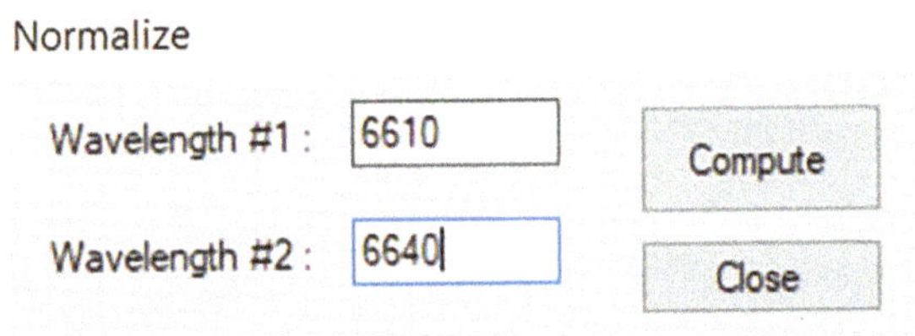

Fig. 2.4.2 — ISIS software normalization window. The values for the normalization zone are to be entered in the wavelength #1 and #2 boxes in ångströms.

Enter in the 2 boxes field, the wavelengths that enclose the area for spectrum normalization with value 1 in intensity. This zone must be free of spectral lines and must be fairly narow, as in the example in Figure 2.4.2.

The continuum of each star is close to the "ideal" Planck curve of a black body at a given temperature. This indicates the temperature at the surface of the observed star. Note that the flux maximum of the hot star's black body is not visible on the graph in Figure 2.4.1, as it is located in the near UV. Similarly, for the profile of the cold star, the maximum is located further out in the near IR. Note also that the molecular lines of cold stars (TiO_2 for example) or the lines of Wolf-Rayet stars may mask the continuum.

Planck's law gives the energy distribution according to wavelength λ of a black body at surface temperature T according to the equation:

$$I_\lambda = \left(\frac{1}{\pi}\right)\left(\frac{C_1}{\lambda^5}\right)\left(\frac{1}{e^{\left(\frac{C_2}{\lambda T}\right)} - 1}\right)$$

Where:
$C_1 = 3.74185 \cdot 10^{-5}$ erg.cm^2.s^{-1}.
$C_2 = 1.43823$ cm.K.
I_λ: the intensity for a given wavelength value λ.
T: the temperature in Kelvin (K).
λ: the wavelength in cm (note the unit).

Wien's law can be used to determine at what wavelength λ_{max} energy is at its maximum as a function of temperature T of the black body:

$$\lambda_{max} = \frac{2\,898}{T}$$

Where:
T: the blackbody temperature in Kelvin.
λ_{max}: the wavelength of maximum energy in µm.

For example, a black body with a temperature of 5 000 K will emit a maximum of energy for λ = 0.5796 µm, or 5 796 Å.

We can present the 1D spectra (obtained after processing) other than a curve, by recreating a synthetic 2D colour spectrum. To do this, we are going to use Visual Spec which makes it easy to generate from a fit 1D file:

1) First, display the spectral profile using the "**Open profile**" from menu "**File**".

2) Next, we use the "**Synthesis**" which can be found in the "**Tools**" menu.

3) A black-and-white synthetic spectrum is displayed and it can be colored by right-clicking on it, then select the function "**Colorer**" from the popup menu.

4) Save the synthetic spectrum by right-clicking again on the spectrum then by selecting "**Exporter**". Give a file name and specify the exported image format (.png, .bmp or .jpg).

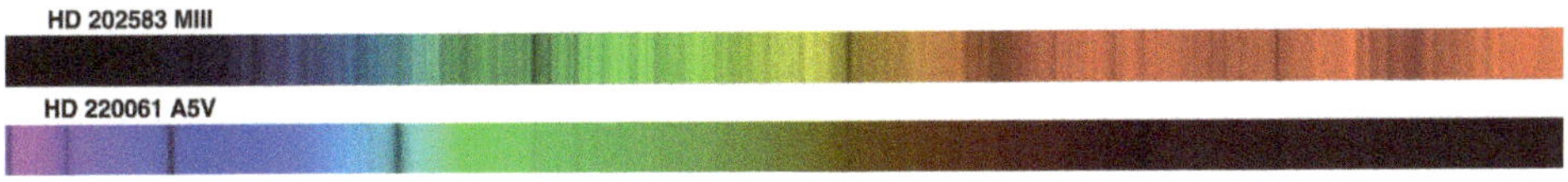

Fig. 2.4.3 — Synthetic spectra. Spectra of a type MIII cold star and a hot A5V-type star generated using Visual Spec.

You can do the same with all your spectra. For the result to be consistent, you need to:

• Correct spectra for instrumental response (a reference star taken at the same altitude in the sky as the targets).

• Normalize each spectrum to a relative intensity of 1 on identical wavelength range for all in a line-free zone close to $H\alpha$. We often use the values 6 610 and 6 640 Å.

Finally, assemble the resulting synthetic spectra into a single image superimposed by decreasing temperature. At the top of the image, we place (for example) the hottest and at the bottom the coldest.

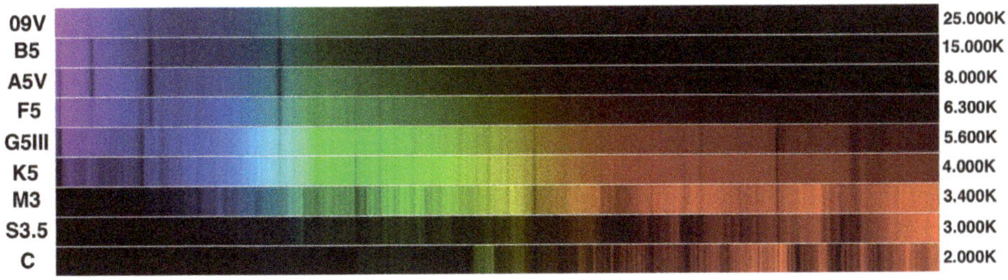

Fig. 2.4.4 — All synthetic spectra. Grouping of the 9 stars used as an example for temperature classification.

2.5 Estimating the temperature of a star

Tools in the Visual Spec and ISIS software packages can be used to estimate the temperature of a stellar spectrum. Similar tools are also available in Matlab or Python libraries to carry out this kind of measurement. Here are two simple methods for estimating temperature with ISIS and Visual Spec.

For a more accurate result, it would be necessary first to "Unredden" the spectrum to take into account interstellar extinction due to the presence of dust and gas on line of sight. Unreddening the lines will be discussed in the Chapter 4 on planetary nebula spectro-imaging.

The instrument response curve, air mass and atmospheric conditions at the time of acquisition are all parameters that can influence temperature measurement by modifying its Planck profile in the blue part of the spectrum.

The spectrum obtained in the visible range is only a small portion of the total spectrum (UV + visible + IR). Without the UV and IR part, the Planck curve is incomplete and we tend to extrapolate the spectral profile according to what we obtain in the visible, which is wrong and too approximate.

In addition, hot stars containing ionized hydrogen and/or helium atoms, do not appear strictly as a black body due to the presence of lines. This leaves us with an estimated temperature rather than an absolute measure but you can still compare two stars using this method.

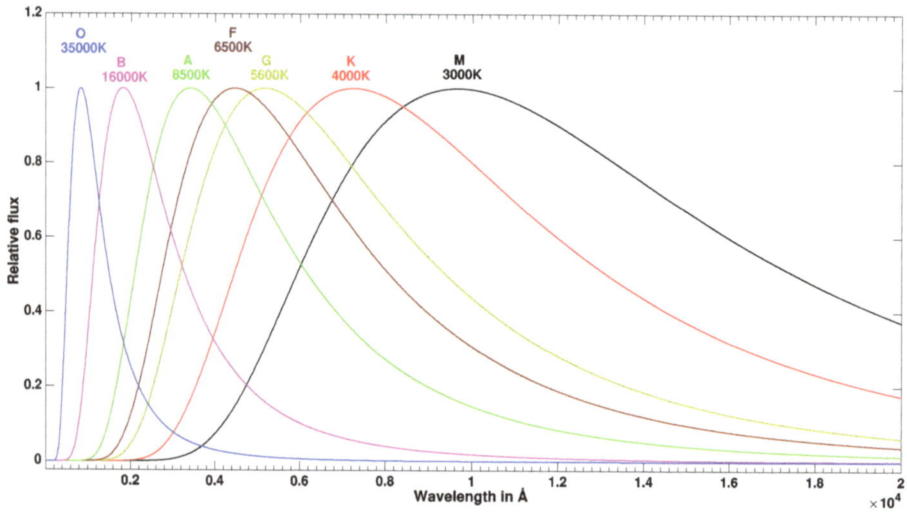

Fig. 2.5.1 — Black body curves. Theoretical curves of black bodies corresponding to stars of type type O to M with their respective temperatures. Curve maxima are not necessarily located in the visible area, only the curves of F, G and K stars are in the visible domain. Generated using ISIS software.

Method 1: Using Visual Spec software

1) First, display the spectral profile using the command "**Open profile**" of the menu "**File**".

2) Then select the "**Planck...**" function from the menu "**Radiometry**".

3) In the window that appears (see Figure 2.5.2), enter a temperature value in kelvin. In the first instance we enter an estimated value, which which we then adjust according to the result obtained.

Fig. 2.5.2 — Planck function window in Visual Spec.

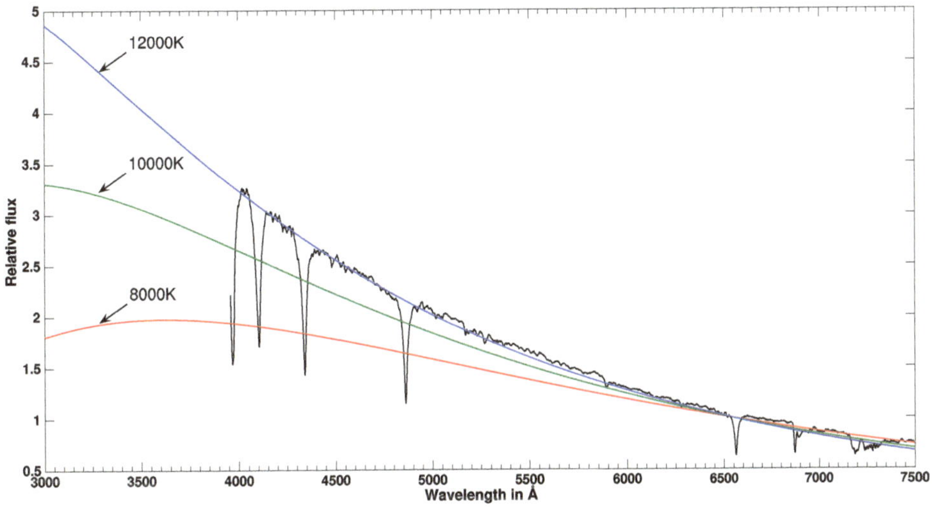

Fig. 2.5.3 — How to determine the right temperature curve. After several tests between 8 000 and 12 000 K, in the end, it is the curve at 12 000 K that superimposes best on the star's spectral profile.

4) The curve of the blackbody whose temperature you have entered is displayed. The curve may be either too high or too low in relation to the to the profile continuum.

We then repeat the points n° 2 and n° 3 until we obtain the value that comes closest to the star's profile. By iteration, we find the best fit to that of the star! (see Figure 2.5.3).

You can also use the function "**Auto Planck...**" where we enter the high and low values in estimated temperature. Visual Spec will search between these 2 values for the curve that best fits the spectral profile.

Method 2: Using ISIS software

ISIS offers a wide range of online orders. Here, we are going to use one of them, the function **l_planck**. Initially, we estimate the approximate temperature of the star under study. For example, for a star of spectral class M, it is pointless asking for a blackbody curve at 10 000 K, we will start with values around 3 000 to 4 000 K.

1) Go to the tab "**Tools**", then "**command line**". Enter the function **l_planck** with the following syntax:

$$l\_planck \quad Name \quad T \quad \lambda_{min} \quad \lambda_{max} \quad P$$

where:

Name: name of the file to be generated, for example "Planck3000K".
T: the blackbody temperature in Kelvin.
λ_{min}: the minimum wavelength in ångströms.
λ_{max}: the maximum wavelength in ångströms.
P: the step in Å between each point on the curve (usually 0.5).

For example, if you wish to generate a 3 000 K curve between 4 000 Å and 8 000 Å, enter the following command (with a space between each argument):

$$l\_planck \quad Planck3000K \quad 3000 \quad 4000 \quad 8000 \quad 0.5$$

2) By validating with "**Enter**", a file is generated "**Planck3000K.fit**" in the current directory used by ISIS.

3) Then compare the star's profile with the file just created in the tab "**5-Profile**" having first normalized the 2 profiles in the same area of the spectrum. To do this, click on the box "**Compare**" and the 2 curves are displayed together.

4) The blackbody curve may be either too high, or too low in relation to the star's continuum. We then adjust the function *l_planck* so as to be as close as possible to

the curve by varying the blackbody temperature.

The two methods described above provide an estimate of the star's temperature with a margin of uncertainty that is nevertheless sufficient for amateur work to classify stars according to their temperatures.

2.6 Bibliography

- Astronomie-Astrophysique, Introduction by Agnès Acker, Dunod, 2005. ISBN 2-10-005462-7.

- Astronomie: méthodes et calculs by Agnès Acker and Carlos Jaschek, Masson, 1995. ISBN 2-225-84689-8.

- Stars and their spectra, An introduction to the Spectral Sequence by James B. Kaler, Cambridge University Press, 2011. ISBN 978-0-521-89954-3.

- Stellar spectral classification by Richard O. Gray, Adam J. Burgasser and Christopher J. Corbally, Princeton University Press, 2009. ISBN 978-0-691-12511-4.

- The Observation and Analysis of Stellar Photospheres by David F. Gray, Press Syndicate of the University of Cambridge.

- Spectral Atlas for Amateur Astronomers by Richard Walker, Cambridge University Press, 2017. ISBN 978-1-107-16590-8.

- Spectrocopie Astronomique by P. Salet, Forgotten Books, 2018. ISBN 978-0-666-48624-0.

- The Classification of Stars by Carlos Jaschek and Mercedes Jaschek, Cambridge University Press, 1987. ISBN 978-0521267731.

- The Spectral Classification of Stars over the last 200, 100, 75 years and in the future by Christopher J. Corbally and Richard O. Gray, Proceedings of the International Astronomical Union 13(S349):489-493;2019.

3. Redshift of galaxies and quasars

In this chapter, we will discover out how to calculate the distance to a galaxy or quasar thanks to spectral line shifts (Doppler-Fizeau effect). A low-resolution instrument will be used to observe weak targets with a preference for LISA, Alpy 600 and UVEX spectrographs (300 lines/mm grating) or Star'Ex LR (*low resolution*).

We can also use a Star Analyser 100 or 200 but if the target is large, it will be difficult to process (no slit), on the other hand, on a faint quasar which is point target, the Star Analyser 100 or 200 can reach very large magnitudes provided the point-like target is well isolated from other stars (no mixing of spectra).

It is also necessary to have a telescope of suitable diameter to collect sufficient light from these targets which can be very faint: A telescope with a diameter of 200 to 300 mm can be a good starting point in this adventure.

Long exposure times are also required, typically between 600 and 1 200 seconds in order to obtain a significant signal across the entire spectrum.

3.1 A brief history

Before the 20th century, astronomers had no precise idea of the distance to certain objects. Catalogs, such as the one done by Charles Messier (1730-1817) of non-point objects, listing 110 objects, does not distinguish between a nebula and a galaxy, he speaks of "diffused objects". He was looking for comets.

From 1908 onwards we begin to distinguish between a galaxy and a nebula in particular through the work of Edward Arthur Fath (1880-1959), an American astronomer who creates spectra on photographic plates with a slitless spectrograph installed on the 0.9 m great refractor at the Lick Observatory in California.

He produced spectra of "spiral nebulae" as he called them at the time. Most of his spectra showed absorption lines, only a handful of objects showed emission lines.

For the first time, in 1917, Vesto Melvin Slipher (1875-1969) measured the radial velocity (towards and awy from Earth) of 25 spiral nebulae using a spectrograph and found that they were moving away from us by measuring line shifts using the Doppler-Fizeau effect. They were redshifted (their light was shifted towards the red of the spectrum), hence the name “redshift” given to this shift noted z.

In 1923, independently of spectroscopy, the works of Edwin Powell Hubble (1889-1953) have made it possible to measure the distance of a galaxy according to the luminosity of certain variable stars (Cepheids), hosted by these galaxies.

In 1927, the Belgian physicist, Georges Lemaître (1894-1966), wrote an article showing that there is a constant relationship between escape velocity and distance.

In 1929, the famous equation linking escape velocity and distance is published as the Hubble Law, it will only become Hubble-Lemaître’s law in 2018. It uses Hubble’s constant (H_0) which has continued to evolve in recent years but whose current consensus value tends towards $H_0 = 74.03$ km.s^{-1}.Mpc^{-1}.

In the years between 1940-43, the American astronomer Carl Keenan Seyfert (1911-1960) discovered only a small number of galaxies (2 to 3 % of the total) having a very luminous core showing a spectrum with emission lines.

Some of these lines are broader than in the spectra of "normal" galaxies. Depending on the type of spectral lines, these galaxies were classified at the time into 2 types:

- **Type 1 Seyfert galaxies**: spectra with both broad and narrow emission lines.

- **Type 2 Seyfert galaxies**: spectra with only narrow emission lines.

Seyfert showed that the width of the lines was due to the Doppler-Fizeau effect but without going any further in investigating the reasons for this difference.

It was only later that it was determined that these lines do not occur in the same area of the galactic nucleus. Broad lines are produced by "permitted" transitions whereas narrow lines are produced by "forbidden" transitions (which can only occur in areas with very few chemical elements so not in the same place as the broad lines). Broad lines are produced close to the accretion disk spinning around the central black hole and fine lines at greater distances.

These 2 types of classification were then completed/subdivided by creating other classes but in the course of observations over the following decades, we realized that this type of classification did not represent all the cases we encountered.

Every active-core galaxy spectrum is almost unique, even within the same class, For example, the spectra of Seyfert type 1 galaxies are not all alike.

Galaxies with this type of very luminous nucleus are grouped under the name "AGN" for *Active Galactic Nuclei.* The luminosity of the core is not only due to the massive presence of thousands of stars, but because some are attracted by the presence of a black hole at its center which excites their gases before sucking them in. Their light becomes very intense and displays a spectrum with certain lines in emission.

In 1985, studying Seyfert type 1 and 2 galaxies in particular, a unified model was proposed for these active galaxi nuclei (AGN).

The publication by these 2 American astrophysicists shows that observations vary depending on the line of sight and therefore the orientation of the galaxy with respect to the observer and, depending on the case, the galaxy can be viewed from the side, from the front, or be hindered by ionized gases masking part of the spectrum of the central zone of the galactic nucleus.

They found by observing in polarized light that NGC 1068 (Seyfert type 2) showed broad stripes, which contradicts the classification of this type of Seyfert type 2 galaxy which in principle should only show narrow lines in emission.

All these situations generate very different spectra.

3.2 Choice of targets

To make redshift measurement easier, it is advisable to focus on galaxies with active cores (AGN) such as Seyfert galaxies, LINER galaxies, quasars and other distant objects showing a spectrum including emission lines, much easier to identify than a spectrum with only absorption lines.

Name	Type	AD (J2000)	DEC (J2000)	B Mag.	V Mag.	R Mag.
M 81	Sy2	09h 55' 33,17"	+69° 03' 55,06"	7,89	6,94	~
NGC 253	SyG	00h 47' 33,13"	-25° 17' 19,68"	8,03	~	6,94
M 106	Sy2	12h 18' 57,62"	+47° 18' 13,39"	9,14	8,41	8,11
M 51	GiP	13h 29' 52,69"	+47° 11' 42,93"	9,26	8,36	8,40
NGC 4945	Sy2	13h 05' 27,27"	-49° 28' 04,44"	9,31	14,40	7,55
M 63	LIN	13h 15' 49,27"	+42° 01' 45,72"	9,34	8,59	8,35
M 87	AGN	12h 30' 49,42"	+12° 23' 28,04"	9,59	8,63	~
M 65	GiP	11h 18' 55,90"	+13° 05' 32,30"	9,60	~	~
M 77	GiP	02h 42' 40,70"	-00° 00' 47,85"	9,61	8,87	10,10
NGC 1097	LIN	02h 46' 19,05"	-30° 16' 29,68"	9,97	9,48	8,72
NGC 1365	Sy1	03h 33' 36,45"	-36° 08' 26,37"	10,08	9,63	8,79
NGC 2841	LIN	09h 22' 02,67"	+50° 58' 35,73"	10,09	9,22	~
M 96	GiP	10h 46' 45,74"	+11° 49' 11,78"	10,15	9,25	8,99
M 61	Sy2	12h 21' 54,92"	+04° 28' 25,59"	10,18	9,65	~
NGC 1566	Sy1	04h 20' 00,39"	-54° 56' 16,62"	10,19	9,73	9,21
M 90	Sy2	12h 36' 49,80"	+13° 09' 46,52"	10,26	9,54	~
NGC 5195	GiP	13h 29' 59,59"	+47° 15' 58,06"	10,45	9,55	9,31
M 58	SyG	12h 37' 43,59"	+11° 49' 05,12"	10,48	9,66	~
NGC 1672	SyG	04h 45' 42,55"	-59° 14' 50,68"	10,53	9,68	9,53
NGC 4125	GiP	12h 08' 06,017"	+65° 10' 26,88"	10,65	9,72	~
NGC 1808	Sy2	05h 07' 42,34"	-37° 30' 46,98"	10,80	9,94	9,36
NGC 2768	SyG	09h 11' 37,50"	+60° 02' 13,95"	10,84	9,87	~
NGC 3077	GiP	10h 03' 19,09"	+68° 44' 01,55"	10,85	10,14	9,74
NGC 6300	Sy2	17h 16' 59,54"	-62° 49' 13,94"	10,88	13,08	9,44
NGC 4450	LIN	12h 28' 29,59"	+17° 05' 05,97"	10,90	10,08	~
NGC 6221	Sy2	16h 52' 46,32"	-59° 13' 00,99"	10,90	13,45	9,68
NGC 7582	GiP	23h 18' 23,62"	-42° 22' 14,06"	10,92	10,62	~
M 109	GiG	11° 57' 35,96"	+53° 22' 29,00"	10,94	~	9,57
M 98	LIN	12h 13' 48,29"	+14° 54' 01,95"	10,95	10,14	~
IC 1459	GiP	22h 57' 10,60"	-36° 27' 43,99"	10,96	11,85	9,34
NGC 7213	Sy1	22h 09' 16,21"	-47° 10' 00,11"	10,97	12,08	10,50
NGC 4216	GiP	12h 15' 54,44"	+13° 08' 57,84"	10,99	10,01	~
NGC 5033	Sy1	13h 13' 27,53"	+36° 35' 37,14"	11,01	12,03	9,73

Fig. 3.2.1 — List of Seyfert galaxies. Extracted from a SIMBAD query.

For example, the spectrum of the M31 nucleus shows only absorption lines in a profile similar to that of a cold cold star, which is to say, having many shifted lines making them harder to identify. We can start with spectra of the most luminous Seyfert galaxies (magnitude V between 7 and 11).

Figure 3.2.1 shows a list of the brightest, extracted from a larger catalog of nearly 950 galaxies in the SIMBAD database:

https://simbad.u-strasbg.fr/simbad/

The second column of Figure 3.2.1 indicates the type of galaxy according to the following abbreviations:
Sy1: type 1 Seyfert galaxies.
Sy2: type 2 Seyfert galaxies.
SyG: Seyfert galaxies (without specifying type 1 or 2).
GiP: galaxies in pairs or in interaction.
AGN: galaxies with active nuclei.
LIN: galaxies whose emission spectrum features broad lines of weakly ionized atoms or LINER galaxies.
We can extract from SIMBAD (https://simbad.u-strasbg.fr/simbad/) more complete lists of objects of the same kind, using precise queries. Here is an example of a list of galaxies with active nuclei.
While searching the internet, this interesting publication was found:

https://ui.adsabs.harvard.edu/abs/1987SoSAO..55.....L/abstract

Which is entitled "A catalogue of Seyfert galaxies". You can copy and paste the reference from this article (Bibcode): **1987SoSAO..55.....L** into SIMBAD in the tab "**Reference query**", then in the area "**Journals**" (Figure 3.2.2).

We then click on "**Submit criteria**" of Figure 3.2.2 and we obtain the following result on Figure 3.2.3.

SIMBAD has found 942 objects. To obtain the list of targets, simply click on the hyperlink "**942**" shown in Figure 3.2.3, the text file generated can then be saved and included in a spreadsheet.

Other queries can be made to search for other types of galaxy galaxies, or even quasars or blazars, using the function "**otype**" associated with the SIMBAD classification list (see Figure 3.2.4).

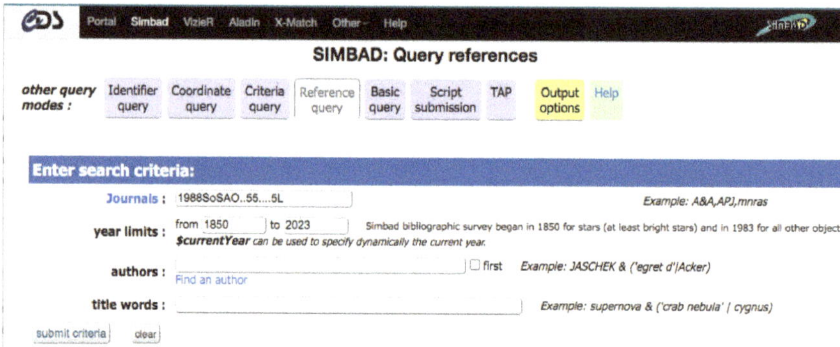

Fig. 3.2.2 — SIMBAD query window. For article references.

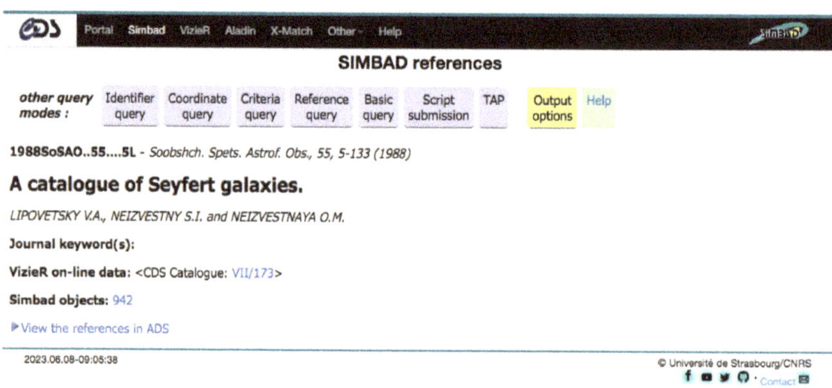

Fig. 3.2.3 — Result of SIMBAD query.

For example, if you want to search for galaxies with active nucleus (AGN) whose magnitude is brighter than magnitude $V = 15$, here is the query to enter in the section "**Criteria query**" of SIMBAD:

otype = "AGN" & Vmag<15

Before sending the query, you can choose the output format in the tab "**Output options**" and also on the right-hand side of the window under the name "**Return**" where you can obtain:

- The total number of objects in the query.
- Or display a list of objects.
- Either display item references linked to this type of object object.

Please note that some requests can take a long time to be processed. You can limit the number of objects by specifying, for example, a limit in magnitude, a limit to the coordinates or by displaying only some of the possible answers. We recommend that you save this list of objects in .csv or .txt format for use in a spreadsheet program.

Type	Description
AGN	Active Galaxy Nucleus
H2G	HII Galaxy
SyG	Seyfert Galaxy
Sy1	Seyfert 1 Galaxy
Sy2	Seyfert 2 Galaxy
SBG	Starburst Galaxy
EmG	Emission-line galaxy
rG	Radio Galaxy
QSO	Quasar
Bla	Blazar
BLL	BL Lac-type blazars
GIP	Galaxy in Pair of Galaxies
LSB	Low Surface Brightness Galaxy
bCG	Blue Compact Galaxy
PaG	Pair of Galaxies
BIC	Brightest Galaxy in a Cluster (BCG)
IG	Interacting Galaxies
PCG	Proto Cluster of Galaxies

Fig. 3.2.4 — List of galaxy classifications according to SIMBAD.

3.3 Making a galaxy spectrum

Now it is time to get hands-on with the spectra of a galaxy or quasar. Remember that you need a low-resolution slit spectrograph like an Alpy 600, a LISA, the Star'Ex LR, a DADOS or a LHIRES III equipped with a grating of 300 lines/mm or with an UVEX also equipped with a grating of 300 lines/mm.

A series of spectra will be needed to measure the radial velocity and to deduce the distance of the target. To start practising, select a fairly bright target of magnitude $V < 11$.

Depending on the type of target, it will be necessary to make the most precise autoguiding possible, to hold the galaxy's nucleus in the slit.

There are two possible situations: the object may be extended (nearby galaxy) or point-like (quasar, blazar, galaxy far away). Depending on the type, acquisition techniques are not quite the same.

Fig. 3.3.1 — Non-point target. The galaxy NGC 7410 (magnitude $V = 11.80$) is centred on the slit of the Alpy 600 spectrograph. Autoguiding is performed on the star indicated by the arrow.

3.3.1 Non-point target

If we are dealing with a non-point target where the shape of the galaxy is clearly visible, the central core must be centred in the spectrograph slit.

The 2D spectrum will show the spectrum of the nucleus as well as those of the arms of the galaxy which will also be present in the spectrograph slit.

If the core is sufficiently bright, we can attempt autoguiding on it, but it is best to use a star in the field that's brighter and more a point-like target than the target.

As far as possible, the spectrograph can also be oriented so that the ellipse of the galaxy's major axis is aligned with the slit, which will give us the spectrum of the entire diameter of the galaxy to measure its rotation velocity, or even to highlight certain HII regions.

During the spectral data reduction process, Make sure you define the binning zone on the galaxy's nucleus. This can be a tricky operation, especially with small focal lengths, as it can come down to the pixel and distort the measurements if the core is exceeded, in this case, part of the rotation velocity of the galaxy's arms impact the final calculations.

You will also need to select a subtraction zone for the background sky. This selection can be problematic for large galaxies that occupy the entire length of the slit. In this case we can make just adjacent to the galaxy, a spectrum of the background sky with the same total exposure time, which is then subtracted from the galaxy spectrum. Of course, this takes longer to achieve.

To begin with, it is best to choose galaxies that are bright and are not too large in relation to the spectrograph slit.

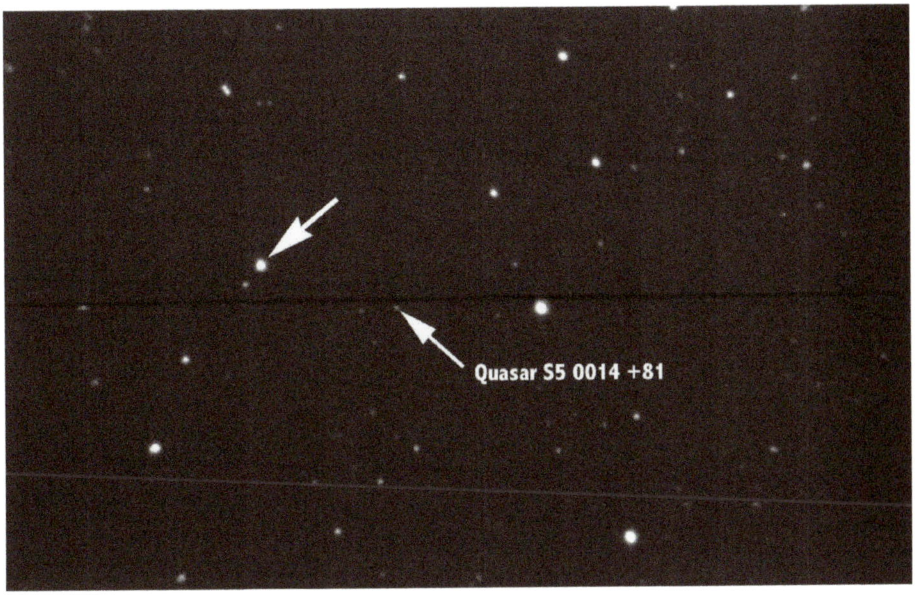

Fig. 3.3.2 — Point target. The Blazar 6C 001403+811827 or S5 0014+81 (magnitude $V = 16.52$) is centred on the spectrograph slit. Autoguiding is performed on the star indicated by the arrow to the left of the target and above the slit.

3.3.2 Point target

If you have a point target such as a quasar or blazar, we proceed in the same way as for a star. But in extreme cases of very faint targets, you will not be able to see them easily with the autoguiding camera except for exposures more than 20-30 seconds, which will allow it to be centred on the slit.

Then we will use a brighter star present in the sensor field to perform autoguiding, taking care not to saturate the sensor and also not to use exposure times < 2 seconds in order to freeze turbulence.

Once the target has been centred on the spectrograph's slit and autoguiding has been launched, the spectral exposures can begin. We will make unit exposures of 900 to 1 200 seconds over a total duration of 2 to 3 hours to obtain a better signal-to-noise ratio. And don't forget to check from time to time that the target is still centred!

3.4 The calculation methods

To calculate the redshift noted "z", we need to measure the position of one or more known lines on the spectrum obtained, which we'll denote by λ_{Obs}. Knowing λ_0, the rest wavelength of these lines, we will use the following equation to calculate the radial velocity V_r:

$$z = \frac{\lambda_{Obs} - \lambda_0}{\lambda_0} = \frac{V_r}{c} = \frac{\Delta\lambda}{\lambda_0}$$

Where:
c: the speed of light (299 792.458 km.s^{-1}).
$\Delta\lambda$: the difference between the observed wavelength λ_{Obs} and the wavelength "at rest" λ_0.

Which gives:

$$V_r = cz$$

The distance to target in light-years, $d_{a.l.}$, is calculated using the following relationship:

$$d_{a.l.} = \left(\frac{V_r}{H_0}\right) 3.26156 \cdot 10^6$$

Where H_0 is Hubble's constant. Over the past few decades, its value has often varied between 50 and 100 km.s^{-1}.Mpc^{-1}. Currently the most recently calculated value (in 2019) is H_0 = 74.03 km.s^{-1}.Mpc^{-1} at $\pm$ 1.42 km.s^{-1}.

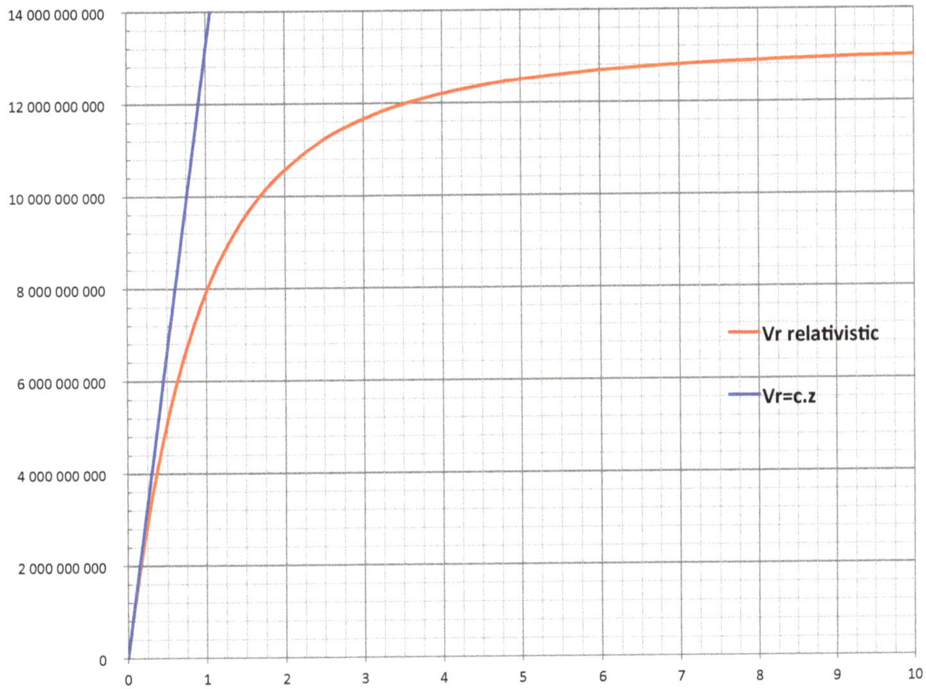

Fig. 3.4.1 — Distance curves in light-years as a function of redshift value z**.** On the x-axis, the redshift z, on the y-axis, the distance in light-years (l.y.) calculated with a Hubble constant value of H_0 = 74.03 km.s^{-1}.Mpc^{-1}. The distance in Mpc is converted into l.y. with 1 Mpc = $3.26156 \cdot 10^6$ light-years.

If you wish to have the result of the distance D in mégaparsecs (1 Mpc = 10^6 pc), the equation then becomes:

$$D = \frac{V_r}{H_0}$$

Remark 3.4.1 To convert D into light-years, the ratio between parsec and light-year is:

- 1 parsec (pc) = 3.26156 light-years (l.y.).
- 1 megaparsec (Mpc) = $3.26156 \cdot 10^6$ light-years (l.y.).

• 1 parsec = $\frac{648\,000}{\pi}$ astronomical units (AU).

• 1 AU: the distance between the Earth and the Sun, whose value officially set by the IAU in 2018 at 149 597 870 700 meters, 8 minutes 19 seconds in light time.

If the target's escape velocity is not negligible compared with the speed of light (< 10 % of c, either 30 000 km.s^{-1}), use the following relativistic equation:

$$Vr_{Rel} = c\frac{(z+1)^2-1}{(z+1)^2+1} = c\frac{\left(\frac{\lambda_{Obs}}{\lambda_0}\right)^2-1}{\left(\frac{\lambda_{Obs}}{\lambda_0}\right)^2+1}$$

From what value of redshift z should the relativistic equation be used? The answer is in the graph in Figure 3.4.1.

The red curve is calculated with the relativistic equation, the blue with the simplified equation. ($Vr = cz$). The graph in Figure 3.4.1 shows that the 2 curves start to diverge from a redshift $z = 0.10$. This must be taken into account when measuring velocity.

Beware also of the value of z, which may be negative for some galaxies in the local group due to have a blue-shifted spectrum. Some examples in the table in Figure 3.4.2 (non-limiting list).

Object	V Mag.	z	Object	V Mag.	z
M31	3.44	-0.001000	M32	8.08	-0.000664
M33	5.72	-0.000598	M110	8.07	-0.000820
IC10	9.5	-0.001160	NGC6822	8.1	-0.000190
NGC185	9.2	-0.000680	NGC147	9.5	-0.000644
UGC12613	12.52	-0.000611	VLM	11.10	-0.000434
LGS3	14.3	-0.000955	IC 1613	10.01	-0.000777
Phoenix	13.2	-0.000043	UGC4879	13.2	-0.000233

Fig. 3.4.2 — Table of nearest galaxies. Negative z redshift values. The V magnitude indicated is according to the SIMBAD database.

3.5 Example with NGC 7469

This target was chosen because it is easily accessible from any observation site with a declination of +8° and quite luminous (magnitude $V = 12.34$). This is a type 1 Seyfert galaxy, which shows both broad and fine emission lines. (The different types of active-nuclei galaxies will be described later in the book).

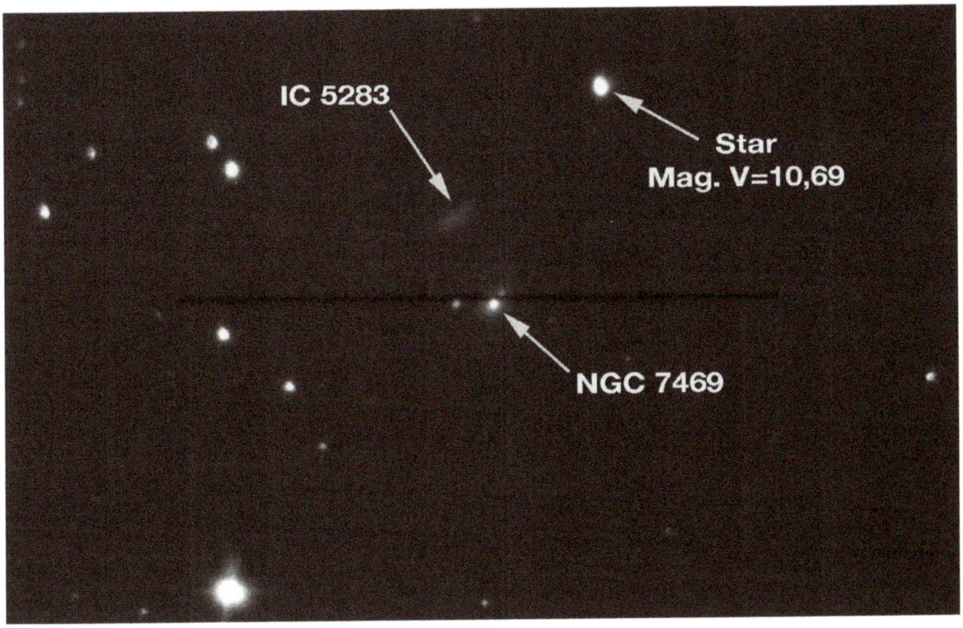

Fig. 3.5.1 — The autoguider sensor field on NGC 7469. The galaxy is centred on the slit, and autoguiding will be performed on the star of magnitude $V = 10.69$ indicated by an arrow.

The target is centred on the slit as shown in the Figure 3.5.1. This example shows that no attempt has been made to orient the spectrograph so as to align the galaxy's long axis with the slit, and it should have been turned through an angle of around 45 degrees. This misalignment will have an impact if we wish to measure the rotation velocity of the galaxy's arms but none if we simply measure its redshift.

However, it is important to position the galaxy nucleus symmetrically in the spectrograph slit and check that the position of the core does not drift during acquisition.

To begin with, we launch the autoguiding on a star close to the target, checking that the core remains centred on the slit. We carry out several unit acquisitions of 900 to 1 200 seconds. Here, 4 exposures of 1 200 seconds taken with an Alpy 600 and an

ATIK 414EX science camera (pixels of 6.45 μm in binning 1x1).

The optics used are a 305 mm Newton telescope open at F/4, which gives the raw spectrum shown in Figure 3.5.2.

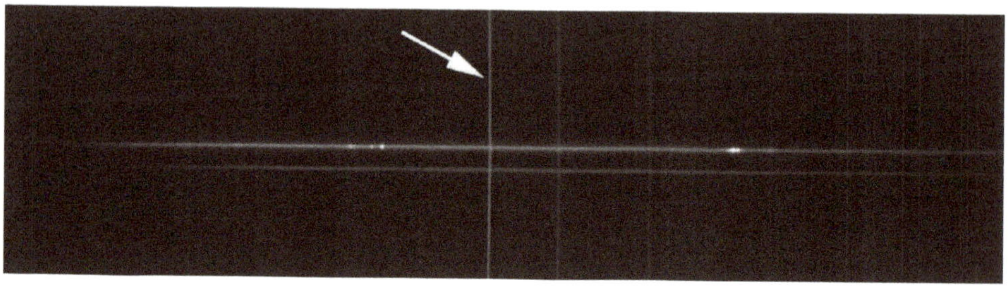

Fig. 3.5.2 — Raw spectrum of NGC 7469. Addition of 4 raw exposures of 1 200 seconds, either 1 h 20 m cumulated. The arrow indicates an atmospheric line of particular interest for recentring the spectral calibration polynomial: [O I] 5 577.35 Å.

3.6 Identifying the spectral lines of the galaxy NGC 7469

Now that the spectra have been created and processed with your favorite software, you will need to identify the lines present. For this particular galaxy, many emission lines are easy to identify using the graph in Figure 3.6.1 and the table in Figure 3.6.2.

This galaxy shows both narrow and broad lines, mainly the Balmer series (hydrogen) typical of a Seyfert type 1 galaxy. We can see that the forbidden line [N II] 6 548 Å is almost confused with the $H\alpha$ line given the low resolution of the spectrum. On other galaxies, it can be clearly seen.

Other lines, such as the [Fe X] 6 374.10 Å and the [Fe VII] 6 087 Å may be present in some spectra of Seyfert type I galaxies but remain fairly weak in intensity. The forbidden line [O II] denoted 1 in the table on Figure 3.6.2 is in fact the mixture of 2 very close lines separated by 2.783 Å (blend) but the one at 3 726.03 Å is much more intense than the second, located at 3 728.81 Å.

The forbidden line [N I] is also a mixture of 2 close lines at 5 200.26 Å and 5 198.52 Å. The table shows the average value of these 2 lines, i.e. 5 199.39 Å.

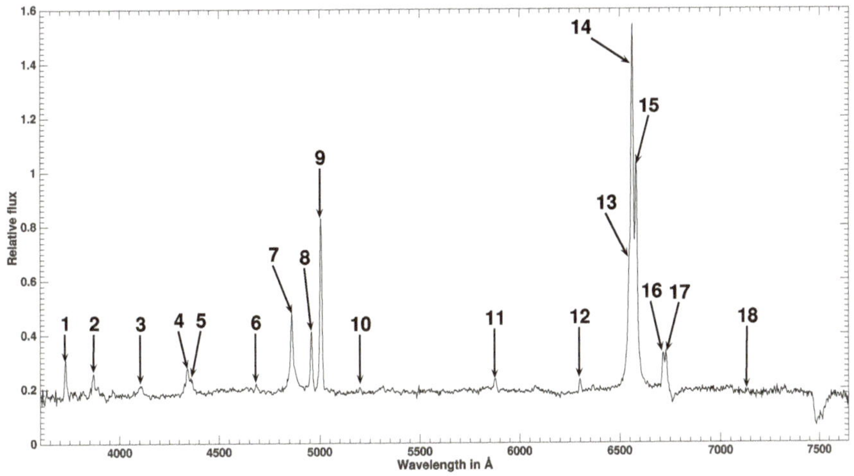

Fig. 3.6.1 — Identification of emission lines in NGC 7469. The graph has been shifted in wavelength by the value of the redshift z calculated for this spectrum: it therefore indicates line values "at rest". On the x-axis, wavelength in ångströms. On the y-axis, intensity relative to the continuum.

N°	Element	Value [Å]	Largeur des raies	N°	Element	Value [Å]	Line width
1	[O II]	3 726.03	Narrow	10	[N I]	5 199.39	Narrow
2	[Ne III]	3 868.76	Narrow	11	He I	5 875.62	Broad
3	Hδ	4 101.74	Broad	12	[O I]	6 300.30	Narrow
4	Hγ	4 340.47	Broad	13	N II]	6 548.06	Narrow
5	[O III]	4 363.21	Narrow	14	Hα	6 562.82	Broad
6	He II	4 685.71	Broad	15	[N II]	6 583.46	Narrow
7	Hβ	4 861.33	Broad	16	[S II]	6 716.44	Narrow
8	[O III]	4 958.91	Narrow	17	[S II]	6 730.81	Narrow
9	[O III]	5 006.84	Narrow	18	[Ar III]	7 135.79	Narrow

Fig. 3.6.2 — Table of NGC 7469 lines and their widths. The numbers on each line correspond to those on the Figure 3.6.1.

3.7 Correcting the heliocentric velocity spectrum

The Earth's rotation around the Sun induces a relative and variable velocity between the object and the observer. Depending on the line of sight, it is necessary to correct the spectrum of this velocity according to the following parameters:

• The observation site coordinates (latitude, longitude).
• The date and time of observation.
• The target coordinates (RA and DEC).

There is nothing complicated about it, as most spectro software packages, such as ISIS, make this task easy. Here is how to do it with ISIS:

Fig. 3.7.1 — Window "Observatory" of ISIS software. Coordinates of the observation site.

First, check the tab "**Settings**" of ISIS software (see Figure 3.7.1), that the coordinates of the observation site are correctly filled in respecting the format in decimal degrees and not in degrees, minutes, seconds.

In the tab "**Misc**", then "**Heliocentric velocity**", specify:

• The target's observation date. The time is indicated by the decimals of a 24 hour day.

• On the next line, enter the name of the target, then click on the "**Simbad**" button which automatically retrieves the object's coordinates from Simbad. Of course, you will need Internet access, but if you are not connected, you can also enter them manually.

Finally, click on "**Go**" to display the results in a window (see Figure 3.7.2). A wide range of information is provided, including the air mass, parallactic angle, target height, etc. at the time of observation. For galaxy redshift calculations, we use the heliocentric velocity (spectral shift), which is more than sufficient for low-resolution corrections.

Heliocentric velocity varies between $\pm$ 30 km.s^{-1} in one year. It is negligible for most low-resolution galaxy redshift measurements because the measurement error with this type of spectrograph is close to this value. But it would be a shame to miss out on this correction, as it's so easy to do and removes a source of measurement error. It is even more important at high resolution.

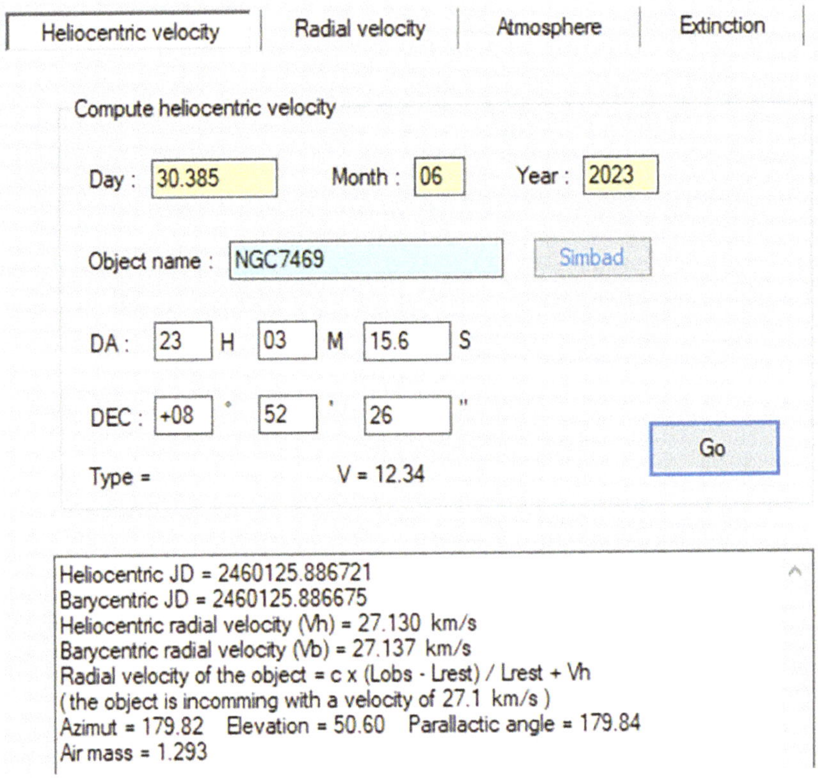

Fig. 3.7.2 — Window "Heliocentric velocity" of ISIS software.

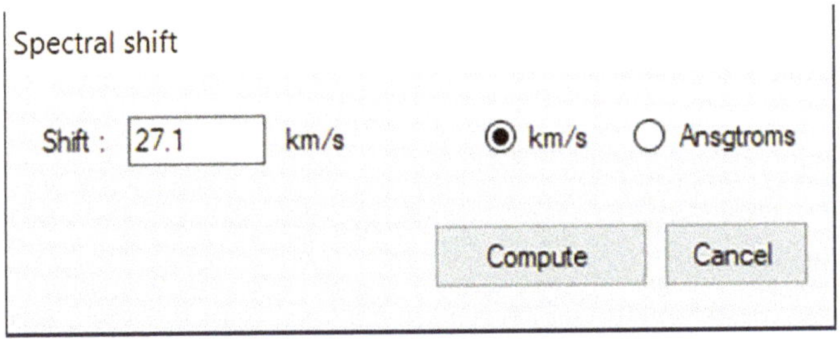

Fig. 3.7.3 — Window "Spectral shift" of ISIS software. With your choice of, a shift in km.s^{-1} or in ångströms.

The heliocentric velocity correction is then applied to the spectral profile from the tab "**5-Profile**", by clicking on the button "**Shift**". Set the heliocentric velocity after checking that the button "**km/s**" is checked (see Figure 3.7.3), then press "**Compute**". The spectral profile is then saved by adding (for example) HVC to the end of the file name so as not to mix spectra corrected or uncorrected for this velocity ("HVC" for heliocentric velocity correction).

3.8 Measuring spectral line positions

A number of free software programs, such as Visual Spec or ISIS, can be used to accurately measure the position of lines in a spectrum. You can also use Python functions if you are familiar with this language, which is widely used in the scientific world and offers many libraries dedicated to astronomy and especially spectroscopy.

Here we describe just one of the methods we use, very easy to set up with Visual Spec software, by modeling a Gaussian curve on a line and measuring its precise position.

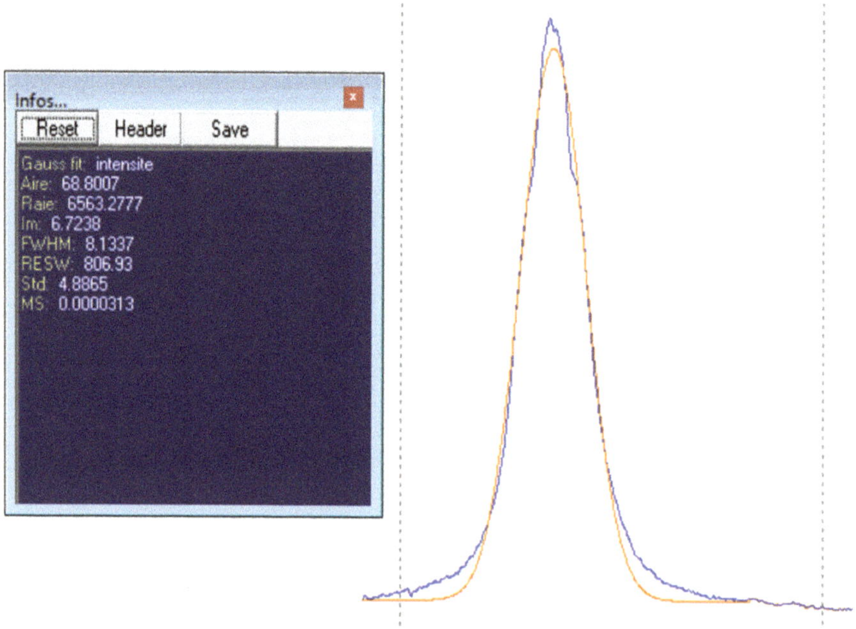

Fig. 3.8.1 — Measuring line positions. With Visual Spec software.

• Display the spectral profile for which you wish to determine the wavelength values of the various lines.

• Select a particular line by moving the mouse while holding down the left mouse button. The selected area is delimited by 2 vertical dotted lines.

• Go to function "**One gaussian fit**" of the menu "**Spectrometry**".

An information window appears with various measurements, including the one we're interested in here: the value of the line position in ångströms. You can check that the Gaussian curve has been modeled by superimposing it on the profile in the graph. (in orange on Figure 3.8.1). Then repeat the same measurement on all the other detectable lines in the spectrum.

A spreadsheet can be filled in with the measured values, as shown in the table below and calculate the redshift for each line, then average the whole.

NGC 7469

	λ o	λ obs	Δλ	z	Vr km/s
[OII]	3726,03	3786,85	60,82	0,016323	4894
[Ne III]	3868,76	3929,84	61,08	0,015788	4733
H δ	4101,74	4168,80	67,06	0,016349	4901
H γ	4340,47	4411,33	70,86	0,016325	4894
[O III]	4363,21	4428,56	65,35	0,014978	4490
He II	4685,71	4759,32	73,61	0,015709	4710
H b	4861,33	4940,53	79,2	0,016292	4884
[O III]	4958,91	5037,53	78,62	0,015854	4753
[O III]	5006,84	5085,77	78,93	0,015764	4726
N I	5199,39	5282,16	82,77	0,015919	4772
He I	5875,62	5970,38	94,76	0,016128	4835
[O I]	6300,30	6401,06	100,76	0,015993	4795
H α	6562,82	6668,93	106,11	0,016168	4847
[N II]	6583,46	6688,75	105,29	0,015993	4795
[S II]	6716,44	6824,84	108,4	0,016140	4839
[S II]	6730,81	6838,62	107,81	0,016017	4802
[Ar III]	7135,90	7252,07	116,17	0,016280	4881
			Z average	**0,016001**	**4797**
			Z (A&A615, A72 (2018))	**0,016268**	**4877**
			Standard deviation	**0,0003371**	**101**

Fig. 3.8.2 — Measurement table of emission lines from NGC 7469. This table can be used to calculate redshift and line velocity as a function of the "at rest" position of these same lines. All these measurements are averaged to determine an average z redshift.

3.9 Major redshifts: $z > 1$ the quasar of "Andromeda parachute"

Like NGC 7469, many galaxies have redshifts of $z < 0.5$. In these cases, it is always possible to identify the lines present even though wavelengths in the visible range may be strongly redshifted or even outside the spectral range.

But as soon as we have high redshifts, it becomes difficult to identify the lines, as their resting values are normally located in the UV at wavelengths not usually encountered in astronomical spectra.

For $z > 2.2$ it is possible to observe as far as the Lymann hydrogen α line located at 1 215.66 Å at rest, but sufficiently shifted to be detected in the visible domain.

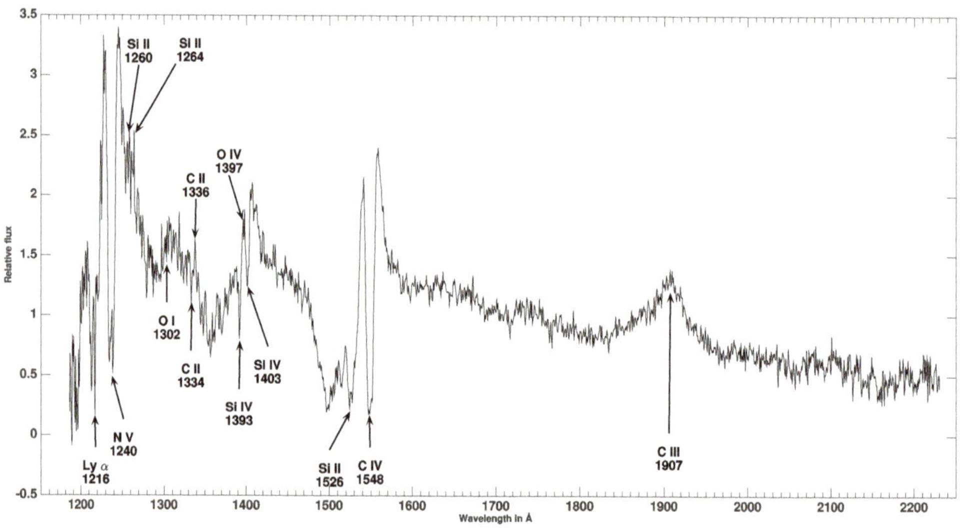

Fig. 3.9.1 — Quasar spectrum J014709+463037 (Andromeda parachute). Realized with a LISA spectrograph (6 x 1 200 seconds of exposure). The spectrum graph has been shifted in wavelength with the calculated redshift value: it therefore shows the values of the lines "at rest". On the x-axis, wavelength in ångströms. On the y-axis, intensity relative to the continuum.

Quasar J014709+463037

	λ o	λ obs	Δλ	z	Vr km/s
Ly α	1216	4075	2859	2,351151	250 768
N V	1240	4145	2905	2,342742	250 541
Si II	1260	4210	2950	2,341270	250 501
Si II	1264	4229	2965	2,345728	250 622
C II	1334	4457	3123	2,341079	250 496
C II	1336	4471	3135	2,346557	250 644
Si IV	1393	4648	3255	2,336683	250 377
Si IV	1403	4680	3277	2,335709	250 350
Si II	1526	5092	3566	2,336828	250 381
C IV	1548	5169	3621	2,339147	250 444
C III	1907	6380	4473	2,345569	250 617

Z average	**2,342042**	**250 522**
Z (A&A, 605 (2017) L8)	**2,341**	**250 494**
Standard deviation	**0,0048557**	**131**

Fig. 3.9.2 — Table of measured quasar redshift values J014709+463037. The reference redshift value $z = 2.341 \pm 0.001$ is taken from a publication in *Astronomy and Astrophysics*, volume 605 (2017) L8.

Let us take the case of the quasar J014709+463037 known as the "Andromeda Parachute". This quasar has a fairly luminous magnitude for such a distance. This is due to a gravitational lens that bends and focuses the light coming from the much more distant object. It was discovered by chance in 1997 in the Pan-STARRS survey.

Its redshift is $z = 2.341$ (a distance of over 11 billion light-years) and its magnitude $V = 15.47$, This makes it a very interesting object to observe spectroscopically, as it reveals unusual lines in the visible range. (hydrogen Lymann α).

We can calculate the redshift of this quasar in the same way as with the NGC 7469 example but using different lines located in the ultraviolet at rest.

Please note! The real velocity must be calculated using the relativistic equation described earlier in this chapter, because the measured escape velocity is far from negligible compared to the speed of light.

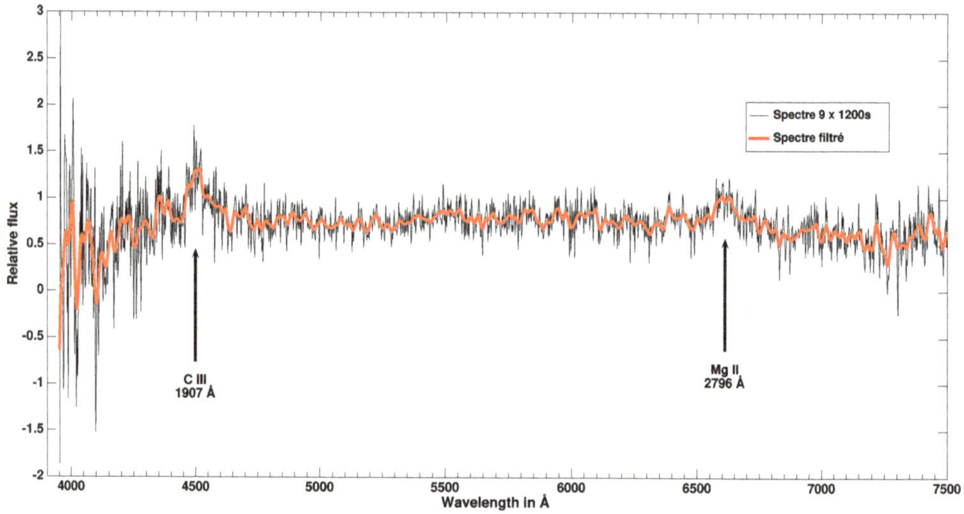

Fig. 3.9.3 — Spectrum of the blazar QSO J1824+1044. Generated with a LISA spectrograph (9 x 1 200 seconds of exposure). The red spectrum represents the filtered spectrum to better identify emission lines. It has been filtered using the "**Filter**" function in the "**5-Profile**" tab of the ISIS software. On the x-axis, wavelength in ångströms. On the y-axis, intensity relative to the continuum.

It is more difficult to clearly identify the lines on a quasar with a redshift between $1 < z < 1.5$ because it shows very few characteristic lines and the magnitude of these targets means there's little chance of getting a good signal-to-noise ratio on the spectrum.

Another example on the spectrum of QSO J1824+1044, it is a blazar of magnitude $V = 17.27$ and a redshift $z = 1.360$ (SIMBAD source). A blazar is seen in the direction of the jet of ejected matter close to the black hole, it has few identifiable lines in the visible range, and in some cases none at all.

Given the low magnitude of this object, the spectrum obtained here with a C14 is very noisy. The only 2 lines that can be identified are:

- Mg II 2 796 Å
- C III 1 907 Å

We will calculate the redshift with just these 2 values, which gives table in Figure 3.9.4.

Blazar QSO J1824+1044

	λ o	λ obs	Δλ	z	Vr km/s
C III	1907	4502	2595	1,360776	208 577
Mg II	2796	6606	3810	1,362661	208 700
			Z Average	**1,362**	**208 638**
			Z SIMBAD	**1,360**	**208 526**
			Standard deviation	**0,001**	**87**

Fig. 3.9.4 — Redshift calculation table for blazar QSO J1824+1044. The reference redshift value was taken from SIMBAD: $z = 1.360$.

Our 2 measurement points nevertheless give a value very close to the value in SIMBAD or other publications (measured $z = 1.362$). This is quite normal, since in this case, the shift of the lines is very large compared with what we were able to calculate with the galaxy NGC 7469 and even if we're less able to identify their position, the measured value is large in relation to the instrumental uncertainty, the measurement is reliable.

For some quasars or blazars, it may be difficult to identify one or more lines if the magnitude is very significant, but also, as mentioned above, because the blazars show almost no lines, because the line of sight of these objects is aligned with the high-energy material jet ejected perpendicular to the accretion disk near the black hole. The emission is essentially in gamma-ray, with hardly any lines in the visible range.

3.10 The case of special galaxies

Some galaxies with active nuclei do not exhibit a spectrum "as shown" in the examples above. NGC 7410, long catalogued as a type 2 Seyfert galaxy, has changed category in a publication by *Astronomy and Astrophysics* (DOI: 10.1051 0004-6361 : 20011252) where it is now referred to as a LINER-type galaxy (*Low Ionization Nuclear Emission-line Region*).

Generate its spectrum: it is a mixture of the absorption lines seen in "classical" galaxies and the very fine emission lines found in Seyfert type 2 galaxies.

In the same way as the previous examples, we can calculate the redshift of NGC 7410 by measuring the lines identified.

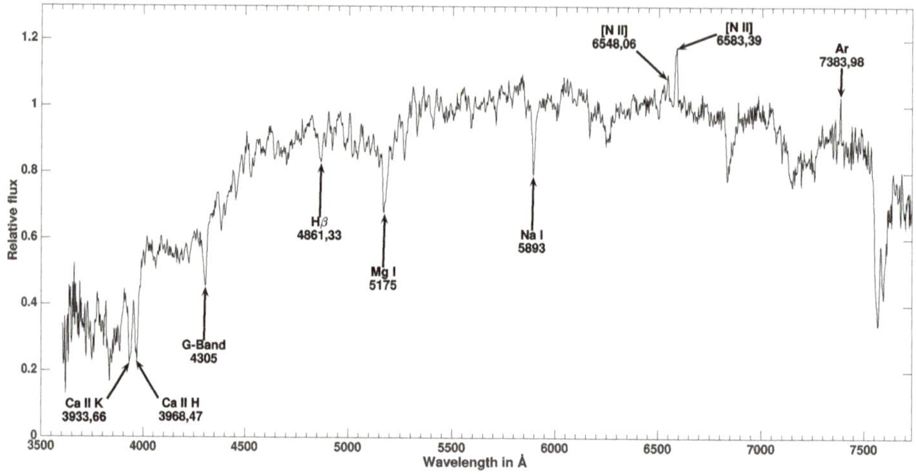

Fig. 3.10.1 — Spectrum of the NGC 7410 galaxy. Produced with an Alpy 600 spectrograph (7 x 1 200 seconds of exposure) from Chile with 2SPOT instrumentation. The graph of the spectrum has been shifted in wavelength by the value of the redshift z calculated on this spectrum: it therefore shows the values of the lines "at rest". On the x-axis, wavelength in ångströms. On the y-axis, intensity relative to the continuum.

NGC 7410

	λ 0	λ obs	Δλ	z	Vr km/s
Ca II K	3933,66	3956,62	22,96	0,005837	1750
Ca II H	3968,47	3988,64	20,17	0,005083	1524
G-Band	4305,00	4327,88	22,88	0,005315	1593
H β	4861,33	4891,93	30,6	0,006295	1887
Mg I	5175,00	5202,99	27,99	0,005409	1621
Na I	5893,00	5927,13	34,13	0,005792	1736
[N II]	6548,06	6581,82	33,76	0,005156	1546
[N II]	6583,39	6622,43	39,04	0,005930	1778
			Z average	**0,005602**	**1679**
			Z SIMBAD	**0,005821**	**1745**
			Standard deviation	**0,000425**	**128**

Fig. 3.10.2 — Table of redshift values for the galaxy NGC 7410. Realized with an Alpy 600 spectrograph (7 x 1 200 seconds of exposure) from Chile. The z redshift reference value was taken from SIMBAD: $z = 0.005821$.

3.11 Find the intruder

To conclude this chapter on calculating galaxy redshifts, here is a little exercise based on a few galaxies in the Markarian chain. This group of galaxies, located almost at the center of the Virgo cluster, is named after the Armenian astronomer Benjamin Markarian (1913-1985) which indicates that this group of 8 galaxies is "animated by a common motion".

The 8 galaxies mentioned by Benjamin Markarian are M84, M86, NGC 4477, NGC 4473, NGC 4461, NGC 4458, NGC 4438 and NGC 4435.

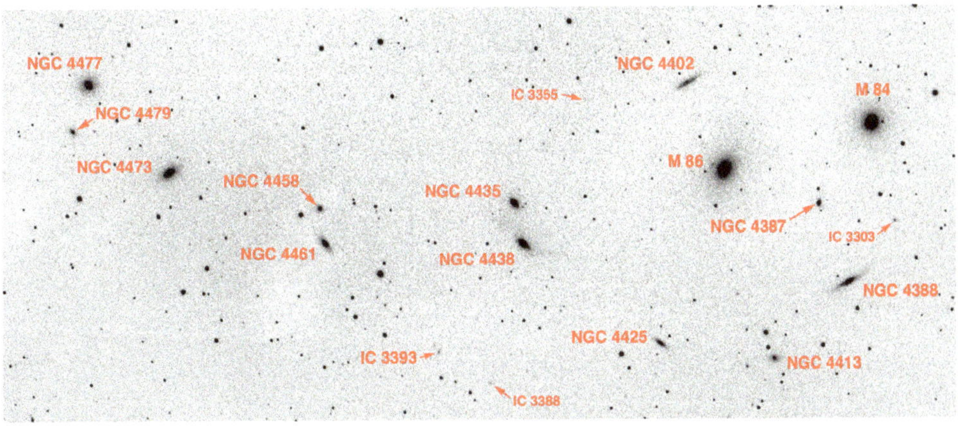

Fig. 3.11.1 — The Markarian galaxy chain. Mosaic created automatically by the Vespera smart telescope (image by François Cochard: 5-hour exposure) with annotations of the main galaxies in this cluster.

Other galaxies are also present in the same area. On Figure 3.11.1, only the brightest are shown. We shall see that Benjamin Markarian's assertion is not entirely valid for the 8 galaxies he mentions.

To do this, start by making a spectrum of the 8 galaxies indicated by Benjamin Markarian. By calculating the redshift of each, we can see that one of them is not moving in the same direction: this is the intruder to unmask! All galaxies have a greater or lesser redshift, and therefore move in the same direction away from the Earth, with the exception of one which moves closer to the observer If you want to find out for yourself, do not look at the table in Figure 3.11.2. The velocity to be measured range between -306 and +2 338 km.s^{-1}. The magnitudes of these galaxies ranges from V = 8.9 to V = 12.07, which makes it easy to observe without too long exposure.

The intruder is the galaxy M86, with a negative redshift and an approach velocity of 306 km.s^{-1}. Two other galaxies (IC 3303 and IC 3355) also show negative redshifts but are not among the 8 described by Benjamin Markarian at the time. Velocity measurements on IC 3355, NGC 4413 and NGC 4438 are at the limit of what a low-resolution spectrograph can detect.

Galaxy	z	Vr	V Mag.	Type
IC 3303	0,00159	476	14,24	Classical
IC 3355	0,00054	162	14,86	Classical
IC 3388	0,005694	1707	14,73	Classical
IC 3393	0,00159	476	14,24	Classical
M 84	0,003392	1017	10,49	Seyfert 2
M 86	**-0,00075**	**-224**	**8,9**	**Classical**
NGC 4387	0,001794	538	12,12	Classical
NGC 4388	0,008419	2524	11,02	Seyfert 2
NGC 4402	0,000791	237	12,6	Classical
NGC 4413	0,00023	69	12,25	Classical
NGC 4425	0,006224	1866	11,83	Classical
NGC 4435	0,002638	791	10,8	Classical
NGC 4438	0,000237	71	10,17	AGN Liner
NGC 4458	0,002258	677	12,07	Classical
NGC 4461	0,006418	1924	11,19	Classical
NGC 4473	0,007485	2244	10,2	Classical
NGC 4477	0,004463	1338	10,42	Seyfert 2
NGC 4479	0,002827	848	12,44	Classical

Fig. 3.11.2 — Table of the main galaxies in this cluster. Galaxies highlighted in yellow are those identified by Benjamin Markarian, other luminous galaxies in this cluster have been added. Highlighted in blue, the intruder's redshift and velocity. Redshift values according to SIMBAD.

3.12 Redshift measurement frame of reference

The measurements we made above were taken between the Earth (possibly corrected for heliocentric velocity, which induces a possible error of ±30 km.s^{-1}) and the target. This is the combined velocity of the Sun and the target. This is the type of measurement that can be found by querying the SIMBAD database but the Sun's rotation around the center of our galaxy induces a velocity of 232.3 km.s^{-1}. This maximum velocity depends on the target's position relative to the Sun's rotation plane around the galaxy's center. The redshift value measured must be corrected by the relative velocity of the Sun on the target's line of sight.

There is an online tool that makes the calculation easy:

http://ned.ipac.caltech.edu/velocity_calculator

Target coordinates and measured redshift are indicated. The value "**Velocity from**" should be on "**Heliocentric**" and the value "**Velocity to**" on "**Galactic Standard of rest**", then click on "**Go**".

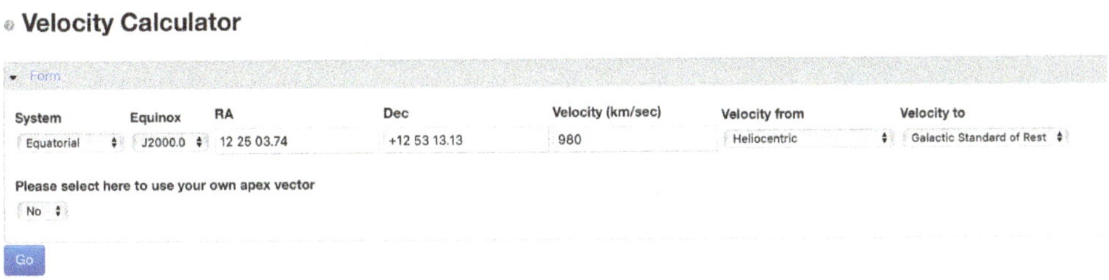

Fig. 3.12.1 — Online tool window "Velocity Calculator".

Taking the example of M84, whose measured redshift is $z = 0.00327$, with an escape velocity of 980 $km.s^{-1}$, the true escape velocity is 925 $km.s^{-1}$. This correction is significant on low redshift galaxies, i.e. galaxies close to us but will be negligible on large redshift galaxies such as quasars or other galaxies with large redshifts ($z > 0.03$).

3.13 Bibliography

- Astronomie-Astrophysique, Introduction by Agnès Acker, Dunod, 2005. ISBN 2-10-005462-7.

- Accurate spectroscopic redshift of the multiply lensed quasar PSOJ0147 from the Pan-STARRS survey, C.H. Lee, Astronomy and Astrophysics, 2017 (doi.org/10.1051/0004-6361/201731695).

- Stellar populations in Seyfert 2 galaxies (doi.org/10.1051/0004-636120011252).

- The Arp atlas of peculiar galaxies by Jeff Kanipe et Dennis Webb, Willmann-Bell, 2007. ISBN 978-0-943396-76-7.

- Astrophysics of Gaseous Nebulae and Active Galactic Nuclei by Donald E. Osterbrock and Gary J. Ferland, University Science Books, 2005. ISBN 1-891389-34-3.

- "Voyage sur les flots de galaxies, Laniakea, et au-delà" by Hélène Courtois, Dunod, 2020. ISBN 978-2-10-080686-7.

- "Galaxies et cosmologie" by Françoise Combes, Misha Haywood, Suzy Collin, Florence Durret and Bruno Guiderdoni, Ellipses, 2009. ISBN 978-2-7298-4083-9.

- AGN surveys by Richard F. Green, Edward Ye Khachikian and David B. Sanders. Astronomical Society of the Pacific conference series, 2002. ISBN 1-58381-127-3.

- Emission lines in active galaxies: New methods and techniques by B.M. Peterson, F.-Z. Cheng and A.S. Wilson, Astronomical Society of the Pacific conference series, 1997. ISBN 1-886733-33-3.

- Evolutionary phenomena in galaxies by J.E. Beckman and B.E.J Pagel, Cambridge University Press, 1989. ISBN 0-521-37193-7.

- The environment and evolution of galaxies by J. Michael and Harley A. Thronson Jr, Kluwer Academic Publishers, 1992. ISBN 0-7923-2541-9.

- The worlds of galaxies II (proceeding of the meeting held in Lyon, France) by Georges Paturel and Chantal Petit, The Astrophysical Letters and Communications, Vol. 31, Numbers 1-6, 1995. ISSN 0888-6512 Alece 7 31 (1-6) 1-362 (1995).

- The Interplay among black holes, stars and ISM in galactic nuclei by Thaisa Storchi-Bergmann, L. C. Ho and H. R. Schmitt, Cambridge University Press, 2004. ISBN 0-521-84803-2.

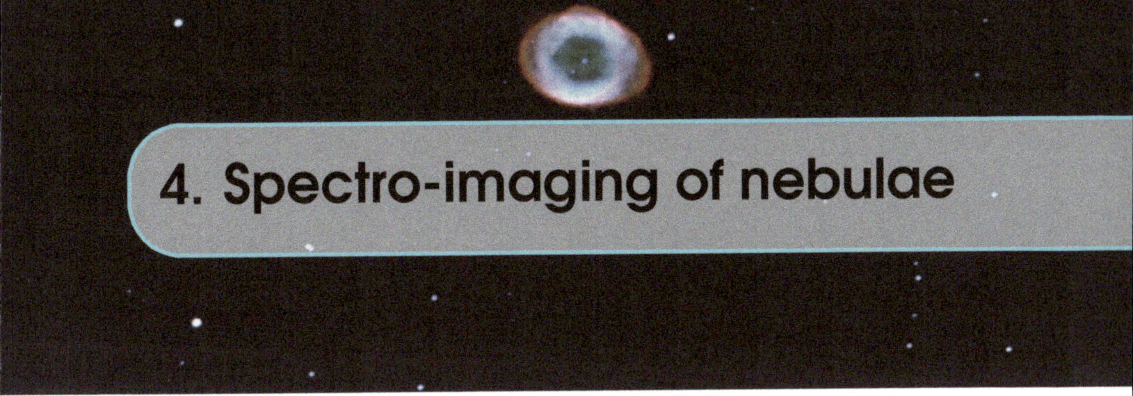

4. Spectro-imaging of nebulae

In this chapter, we will discover how to image a planetary nebula using low-resolution spectroscopy, in the wavelength of your choice, and then how to perform some calculations of physical quantities of the nebula.

The advantage of using a spectrograph is that you can produce a spectrum that can be adapted to a multitude of virtual narrow-band filters.

You can then choose the center wavelength and width of each virtual filter. This allows us to generate images at certain wavelengths of emission lines visible in the spectrum, to construct a resulting color image or to perform calculations to determine the temperature, the electron density or ionization structure of the planetary nebula and other measurements.

This type of experiment can be carried out with a low-resolution spectrograph to cover the entire visible range: from 400 to 700 nm in a single shot. We can use slit spectrographs such as an Alpy 600, the LISA, the Star'Ex or the LHIRES III equipped with a 300 lines/mm grating.

The optical focal length (telescope or refractor) will be decisive, depending on the target and you would not observe the same objects, depending on whether you are using a 80 mm refractor or a 406 mm telescope.

But by choosing carefully the target according to your hardware configuration, you will be able to produce images centered on a specific wavelength.

Low focal lengths are suited to planetary nebulae of several arcminutes in diameter, while we would reserve the smallest nebulae for telescopes with long focal lengths.

We can also calculate the angle seen by the slit for a given configuration. This helps in the choice of targets.

4.1 Principles of spectro-imaging

The principle of reconstructing an image based on spectra consists of "scanning" the planetary nebula by moving the spectrograph's slit slightly after making one (or more) spectra for each position. These data are then used to reconstruct an image based on a selected wavelength, a kind of selective and adjustable filter.

The same principle applies when scanning a document or by using a photocopier. Depending on the size of the planetary nebula and its magnitude, spectra acquisition times can be quite long.

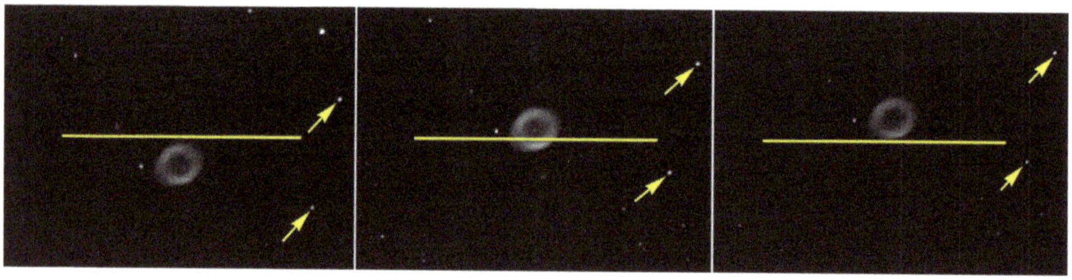

Fig. 4.1.1 — Principle of planetary nebula digitization. Example of a vertical scan from top to bottom using a LISA spectrograph on M57. Stars that can be used as autoguiding setpoints are indicated by an arrow.

Depending on whether the slit is vertical or horizontal, you place the target above or below, to the right or left. The important thing is to ensure that the nebula does not extend past either end of the slit during exposure. We can do a trial sweep to confirm, where it is widest.

Remember also to leave a margin around the object, starting and ending at some distance from the edge of the nebula. Caution! The direction of movement is not important, but it can reverse the images! All you have to do is flip the final image.

Then select a guide star from the autoguiding sensor field. It must not pass through the slit during the scanning session otherwise you'll be unable to autoguide on that guide star.

This star must therefore be chosen carefully before launching the autoguiding. We may try to change the guide star during the acquisition process, but this can lead to shifts when recomposing the image. This should be avoided!

Once the guide star has been selected, we note its coordinates X and Y in pixels that we'll tell the software to execute the autoguiding setpoint at this point.

The methodology for digitizing a nebula is as follows:

1) Position the target next to the slit.

2) Select an autoguiding star that will not pass through the slit during scanning and note its X and Y pixel coordinates.

3) Start autoguiding on the setpoint X, Y.

4) Start acquisition of one or more spectra (depending on target magnitude).

5) After an exposure or a series of exposures on a position (a series might be necessary to obtain a desired signal to noise ratio), shift the autoguiding setpoint by 1 pixel each time: $Y_2 = Y_1 + 1$ pixel or $Y_2 = Y_1 - 1$ pixel depending on whether the target is above or below the slit. (Or $X \pm 1$ if moved horizontally.)

6) Restart acquisitions then repeat the operation until the target emerges from the slit by a few pixels (the margin).

4.2 Target selection

To help select with the selection of a target, Figure 4.2.1 offers you a list of a hundred of the most luminous planetary nebulae, from Messier, NGC (*New General Catalogue*) and IC (*Index Catalogue*) catalogs using the SIMBAD tool from the Strasbourg astronomical data center.

PN	AD J2000	DEC J2000	Size	Mag. Surf.	Mag. Star.
NGC 40	00 13 01.01	+72 31 19.03	1,07' x 0,81'	10,5	11,63
NGC 246	00 47 03.34	-11 52 18.97	4,31' x 3,5'	8,5	12,18
M76	01 42 19.65	+51 34 31.53	4,05' x 1,85'	11	17,48
IC 289	03 10 19.30	+61 19 00.91	0,73' x 0,68'	13	15,9
NGC 1360	03 33 14.64	-25 52 17.97	7' x 5,1'	9,5	11,16
IC 351	03 47 32.98	+35 02 48.63	0,28' x 0,25	12,3	15,7
IC 2003	03 56 21.98	+33 52 30.75	0,27' x 0,29'	11,87	15
NGC 1501	04 06 59.39	+60 55 14.27	1' x 0,96'	12,5	15,17
NGC 1514	04 09 16.98	+30 46 33.47	3' x 2,87'	10,5	9,48
NGC 1535	04 14 15.76	-12 44 21.93	0,95' x 0,94'	9,5	12,2
IC 418	05 27 28.20	-12 41 50.28	1,83' x 1,78'	10	10,12
NGC 2022	05 42 06.19	+09 05 10.58	0,55' x 0,5'	11,5	15
IC 2149	05 56 23.90	+46 06 17.20	0,72' x 0,42'	11	11,3
IC 2165	06 21 42.78	-12 59 14.04	0,32' x 0,32'	12,1	17,9
NGC 2346	07 09 22.52	-00 48 23.61	2,2' x 1,08'	10,5	11,58
NGC 2371	07 25 34.68	+29 29 26.43	2,27' x 3,28'	11,5	14,48
NGC 2392	07 29 10.76	+20 54 42.47	0,9' x 0,9'	9,68	10,5
NGC 2438	07 41 50.51	-14 44 07.47	1,21' x 1,18'	10,5	17,5
NGC 2440	07 41 55.40	-18 12 32.00	1,31' x 1,21'	10,5	17,7
NGC 2452	07 47 26.26	-27 20 06.81	0,83' x 0,46'	12,2	17,3
IC 2448	09 07 06.31	-69 56 30.69	0,46' x 0,46'	11,1	14,25
NGC 2792	09 12 26.56	-42 25 38.92	0,39' x 0,39'	12,5	13,8
NGC 2818A	09 16 10.60	-36 37 26.00	1,68' x 0,82'	12,5	19,4
NGC 2867	09 21 25.38	-58 18 40.62	0,95' x 0,95'	9,5	16,65
NGC 2899	09 27 03.02	-56 06 21.10	2,17' x 1,15'	12,4	15,92
IC 2501	09 38 47.19	-60 05 30.80	0,33' X 0,33'	10,5	14,5
NGC 3132	10 07 01.76	-40 26 11.13	1,62' x 1,10'	8,5	10,01
IC 2553	10 09 20.86	-62 36 48.54	0,41' x 0,41'	11,9	15,5
NGC 3195	10 09 20.91	-80 51 30.73	0,75' x 0,66'	11,5	15,3
NGC 3211	10 17 50.54	-62 40 12.10	0,36' x 0,34'	11,5	19
NGC 3242	10 24 46.13	-18 38 32.29	1,35'" x 1,35'	9,5	12,3
IC 2621	11 00 20.02	-65 14 57.82	0,23' x 0,23'	11,42	15,4
M 97	11 14 47.71	+55 01 08.48	3,4' x 3,4'	11	15,77
NGC 3699	11 27 57.83	-59 57 28.45	2,10' x 1,51'	11	13
NGC 3918	11 50 17.77	-57 10 57.01	1,02' x 1,02'	8,1	14,6
NGC 4071	12 04 14.70	-67 18 36.45	1,33' x 1,02'	13	14
NGC 4361	12 24 30.75	-18 47 05.56	2,11' x 1,82'	10,5	13,2
IC 3568	12 33 06.85	+82 33 50.20	0,4' x 0,4'	11,5	13,5
IC 4191	13 08 47.30	-67 38 37.54	0,47' x 0,36'	11,61	16,35
NGC 5189	13 33 32.87	-65 58 27.04	3,30' x 2,10'	10	14,9

PN	AD J2000	DEC J2000	Size	Mag. Surf.	Mag. Star.
NGC 5307	13 51 03.25	-51 12 20.83	0,64' x 0,4'	11,5	14,7
NGC 5315	13 53 56.97	-66 30 50.85	0,5' x 0,42'	11,5	14,4
IC 4406	14 22 26.14	-44 09 02.18	1,63' x 0,60'	10,5	17,1
NGC 5844	15 10 40.60	-64 40 25.00	1,83' x 1,15'	12,7	?
NGC 5873	15 12 50.86	-38 07 31.58	0,24' x 0,24'	11,6	15,5
NGC 5882	15 16 49.95	-45 38 58.61	0,66' x 0,6'	10	13,4
NGC 5979	15 47 41.17	-61 13 05.54	0,48' x 0,42'	12,1	15,3
PN Sp 1	15 51 40.95	-51 31 28.54	1,5' x 1,5'	13,98	14
NGC 6026	16 01 21.12	-34 32 35.78	0,86' X 0,83'	12,9	?
NGC 6058	16 04 26.54	+40 40 58.94	0,59' x 0,46'	13	13,9
IC 4593	16 11 44.54	+12 04 17.03	0,6' x 0,6'	10,84	11,2
NGC 6072	16 12 58.36	-36 13 47.39	1,77' x 1,33'	13	?
IC 4599	16 19 23.20	-42 15 36.71	0,45' x 0,45'	12,8	?
NGC 6153	16 31 30.57	-40 15 12.64	1,3' x 1,05'	10,7	15,55
NGC 6210	16 44 29.51	+23 47 59.42	1' x 1'	9,9	12,7
IC 4634	17 01 33.58	-21 49 33.01	0,38' x 0,38'	11,3	13,9
IC 4637	17 05 10.47	-40 53 07.70	0,4' x 0,43'	12,5	9,2
IC 4642	17 11 45.20	-55 24 01.66	0,33' x 0,33'	13	?
NGC 6302	17 13 44.50	-37 06 11.00	3,37' x 1,6'	12	7
NGC 6309	17 14 04.30	-12 54 37.89	0,96' x 0,45'	11	14,4
NGC 6326	17 20 46.25	-51 45 15.17	0,54' x 0,37'	11,5	13,5
NGC 6337	17 22 15.67	-38 29 01.73	1,55' x 1,08'	12	14,9
NGC 6369	17 29 20.45	-23 45 34.77	1' x 0,76'	11,5	15,9
IC 4663	17 45 28.62	-44 54 17.98	0,33' x 0,25'	12,5	15,2
NGC 6439	17 48 19.82	-16 28 44.42	0,23' x 0,23'	12,6	15,48
NGC 6445	17 49 15.02	-20 00 34.15	3' x 2,21'	12	19
IC 4670	17 55 07.03	-21 44 40.04	0,22' x 0,22'	13,28	?
NGC 6543	17 58 33.40	+66 37 58.75	1,22' x 1,28'	8,5	11,1
IC 4673	18 03 18.45	-27 06 22.71	0,37' x 0,22'	13	17,6
NGC 6537	18 05 13.10	-19 50 34.88	2' x 0,86'	12	18,8
NGC 6565	18 11 52.48	-28 10 42.42	0,26' x 0,26'	12,5	18,5
NGC 6563	18 12 02.50	-33 52 07.00	0,97' x 0,73'	12,5	18
NGC 6572	18 12 06.31	+06 51 13.02	0,9' x 0,9'	8,5	13,6
NGC 6567	18 13 45.16	-19 04 34.07	0,25' x 0,25'	10,9	14,4
IC 4699	18 18 32.04	-45 59 01.73	0,25' x 0,25'	12,5	15,1
NGC 6620	18 22 54.18	-26 49 17.19	0,18' x 0,18'	12,7	?
NGC 6629	18 25 42.45	-23 12 10.55	0,62' x 0,46'	11,5	12,9
NGC 6644	18 32 34.70	-25 07 44.15	0,25' x 0,25'	11,5	15,6
IC 4776	18 45 50.70	-33 20 34.12	0,3' x 0,3'	11,5	14,1
M 57	18 53 35.09	+33 01 44.88	2,4' x 2,27'	9	15,4

PN	AD J2000	DEC J2000	Size	Mag. Surf.	Mag. Star.
NGC 6741	19 02 37.10	-00 26 56.70	0,27' x 0,27'	11	14,7
NGC 6751	19 05 55.53	-05 59 32.28	0,5' x 0,45'	12	15,48
IC 4846	19 16 28.22	-09 02 36.67	0,18' x 0,18'	12,5	15,2
IC 1297	19 17 23.44	-39 36 46.25	0,52' x 0,37'	10,5	14,2
NGC 6778	19 18 25.00	-01 35 47.00	0,62' x 0,43'	12,3	?
NGC 6781	19 18 28.08	+06 32 19.29	0,9' x 0,9'	11,8	16,8
NGC 6790	19 22 56.96	+01 30 46.46	0,44' x 0,44'	10,45	13,5
NGC 6803	19 31 16.48	+10 03 21.62	0,25' x 0,25'	11,6	15,2
NGC 6804	19 31 35.14	+09 13 31.29	1' x 1'	12	14,4
NGC 6807	19 34 33.53	+05 41 02.50	0,25' x 0,25'	12	?
NGC 6818	19 43 58.02	-14 09 13.44	0,65' x 062'	9,3	16,9
NGC 6826	19 44 48.15	+50 31 30.24	2,1' x 2,1'	9,35	10,4
NGC 6842	19 55 02.10	+29 17 22.00	0,95' x 0,95'	13	16
M 27	19 59 36.36	+22 43 16.31	8,08' x 5,66'	7,5	13,9
NGC 6852	20 00 39.21	+01 43 40.92	0,48' x 0,43'	13	17,9
NGC 6884	20 10 23.64	+46 27 39.63	0,25' x 0,25'	12,5	15,6
NGC 6879	20 10 26.69	+16 55 21.27	0,2' x 0,2'	13	14,8
NGC 6886	20 12 42.83	+19 59 22.63	0,25' x 0,22'	11,8	15,7
NGC 6891	20 15 08.84	+12 42 15.58	0,38' x 0,38'	11	12,4
IC 4997	20 20 08.75	+16 43 53.69	0,25' x 0,25'	11,5	14,4
NGC 6905	20 22 22.99	+20 06 16.28	1,33' x 0,62'	11,5	15,7
NGC 7008	21 00 32.81	+54 32 35.37	1,62' x 1,45'	12	13,2
NGC 7009	21 04 10.81	-11 21 48.58	1,6' x 1,45'	8	12,8
NGC 7026	21 06 18.57	+47 51 06.77	0,77' x 0,35'	12	14,2
NGC 7027	21 07 01.57	+42 14 10.47	1,2' x 1,2'	9	16,3
NGC 7048	21 14 15.18	+46 17 17.10	1,2' x 0,95'	11,5	19,1
IC 5117	21 32 30.96	+44 35 47.54	0,15' x 0,15'	12,5	16,7
NGC 7094	21 36 52.96	+12 47 19.09	1,63' x 1,47'	13,4	?
IC 5148	21 59 35.08	-39 23 08.23	2,27' x 2,15'	12	16,5
IC 5217	22 23 55.71	+50 58 00.46	0,42' x 0,32'	12	15,5
NGC 7293	22 29 38.54	-20 50 13.74	20' x 14'	7	13,5
NGC 7354	22 40 19.87	+61 17 08.74	0,62' x 0,62'	12,5	16,2
NGC 7662	23 25 53.83	+42 32 05.83	1,8' x 1,68'	9	13,2

Fig. 4.2.1 — List of the most luminous planetary nebulae.

The table columns in Figure 4.2.1 are:

- PN: name of the planetary nebula in the Messier, NGC and IC catalogues.

- AD J2000: J2000 coordinates in right ascension, in hours, minutes, seconds (HH MM SS.SS).

- DEC J2000: J2000 coordinates in declination, in degrees, minutes, seconds (DD MM SS.SS).

- Size: dimensions of the planetary nebula in arcminutes along the long and short axes of the object, determined by Aladin software from an image of DSS2 (*Digital Sky Survey*).

- Mag. Surf.: the estimated surface magnitude *V* of the planetary nebula. This is an approximation as it is quite difficult to give exact values given the large number of intensity nuances possible on the various zones of these objects, depending on their morphology.

- Mag. Star.: the magnitude of the central star, if known. The magnitudes were taken from various publications. Here again, these are not necessarily absolute values.

4.3 Practical case on M57

For our first test, we will choose M57, the ring nebula in the Lyra constellation, near Vega. It is a good compromise between a target that is bright enough and a surface that is compact enough not to spend too much time scanning it.

It is possible to estimate the time needed to obtain the complete spectral sequence:

Length to be scanned in pixels + margin = *N* (number of spectra to be produced).
Length of acquisition: *N* x exposure time (single or multiple images).

Depending on target magnitude, you will obtain, for each slit position, 1, 2 or 3 exposures. The more exposures you make, the better the signal-to-noise ratio of the nebula scan, but it will take longer to scan everything.

We must also add the travel time between each shot, neon calibration images (taken regularly to account for any flexure of the spectrograph), images of the reference star and the flats.

Digitizing an object can be very time-consuming and tedious, Fortunately, scripts can help automate acquisition! You can spend a whole night or more there, depending on the target's size, magnitude and declination.

As noted earlier, it is advisable to select the target according to the focal length of your telescope. An essential point is that the nebula does not extend beyond the ends of the slit. If possible, nebular line-free margins for background sky subtraction during data reduction.

In this example, the telescope used is a C14 equipped with a focal reducer that reduces the native aperture ratio from F/D 11 to F/D 5.5. The spectrograph is a LISA equipped with an ATIK 414EX camera used in 1x1 binning. A unit exposure time of 600 seconds is used.

Short focal length instruments are preferable for large planetary nebulae, as you can digitize larger objects more easily with a short focal length because the flux is spread over a smaller area. Feel free to experiment with different nebulae depending on your hardware configuration.

Unit time exposure: start with a simple exposure of few minutes to check the maximum number of ADUs obtained, and then adapt the duration to suit these results.

Be very careful not to saturate very intense emission lines [O III], [N II] and $H\alpha$ aim not to exceed 2/3 of the sensor dynamic range. Depending on the magnitude of the target, you will obtain one or more exposures for each slit position. Increasing the number of exposures improves the signal-to-noise ratio, but it takes longer!

You can also plan to carry out the acquisitions over several nights, remembering to note the exact position of the autoguiding star (its X, Y coordinates on the autoguiding sensor) in order to take over the acquisitions exactly at the position of the last exposure. Of course, the equipment must not be modified between sessions.

Once the acquisitions have been made, do not forget to observe a reference star close to the target. In the case of M57, we can use Vega (AOV), located just next door. Do not forget the series of flats and the calibration lamp spectra (neon or argon/neon). We assume darks and biases are already in stock.

All of the above may seem a little complex at first, but you are invited to try it for yourself, and you will soon see that it is not that difficult once the rest of the system is fully operational (setting up the mount, focusing the target on the spectrograph slit, autoguiding management). Admittedly, this requires a few technical prerequisites, but digitization itself is simple to achieve and no more complicated than making an image through several filters. It can be automated with certain acquisition software with scripting functions.

Now you have all the data you need to extract the image, so let's see how to process it.

4.4 Spectral sequence processing

Let us take the example of processing with ISIS software, you can also do it with other programs, or even create your own Python script. The advantage of ISIS is that all the functions already exist for reconstructing an image from a series of spectra.

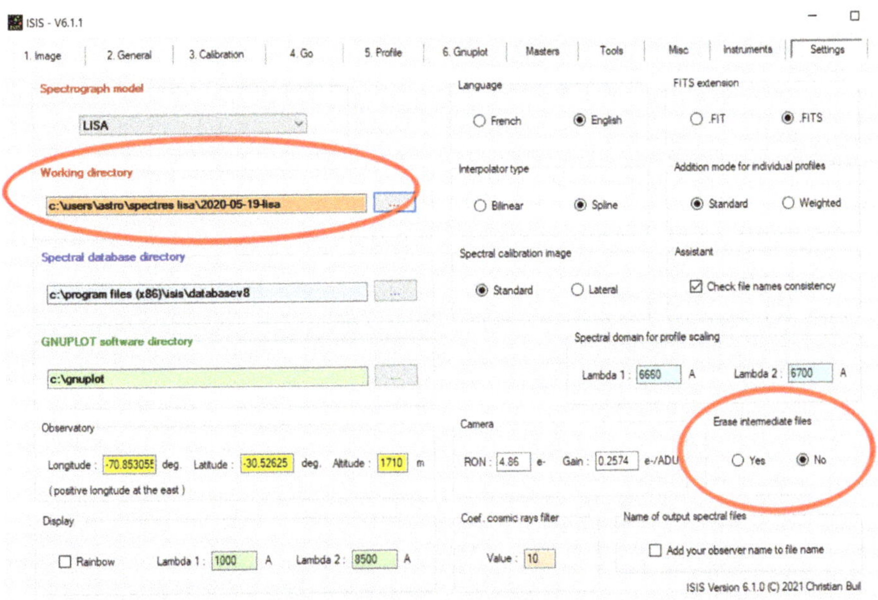

Fig. 4.4.1 — "Settings" tab of ISIS software.

First of all, in order to respect the syntax of file names in ISIS software, your various spectra must be numbered as follows:

M57-1.fits, M57-2.fits, M57-3.fits......., M57-10.fits, ...M57-N.fits

Note: no leading zeroes between the "-" and number. ISIS quite fussy with this regard.

The processing will be described in detail, on the assumption that a spectral image has already been produced for each position of the slit on the nebula.

If you have made several spectra for each position, then you will first have to process each batch of images independently for a given slit position, which can be quite time-consuming, but is guaranteed to give a better signal-to-noise ratio on the final images.

As a first step, we can process a single M57 spectrum in the usual way, calibrating it in wavelength, which will facilitate the subsequent identification of the various lines.

Then apply the following procedure to produce the spectra used to reconstruct the image.

When ISIS is opened, the first thing to do is to specify the working directory containing all the acquisition files. Click on the tab "**Setting**" (see Figure 4.4.1). Select path for working directory. You must also select the option "**No**" for "**Erase intermediate files**" because it is with these "intermediate" files generated by ISIS that we will be able to reconstruct our image. By default, they are destroyed after calculation to keep hard disks from getting cluttered! These files have the generic name @ - : @-1, @-2, @-3... @-n.

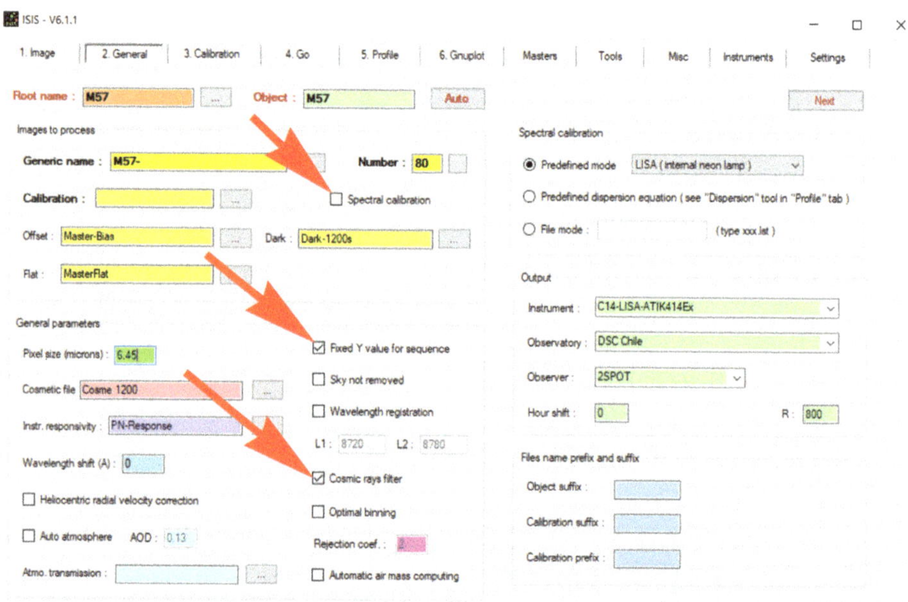

Fig. 4.4.2 — Tab "2-General" of ISIS software.

Then, in the "**2-General**" tab, modify the 3 boxes indicated by a red arrow on Figure 4.4.2:

1) Uncheck the box "**Spectral calibration**" which is of no use here, as we want to reconstruct an image.

2) Check "**Fixed Y value for sequence**", since the spectrum is extended and not a point, we need to tell ISIS to always take the same *Y* coordinate of the image of the nebula center to avoid displacement of the subsequent spectra.

3) Check if necessary "**Cosmic rays filter**" if you see any in your images.

The other boxes are to be completed in the usual way (target name, master file names dark, flat, bias, pixel size of your camera, etc.).

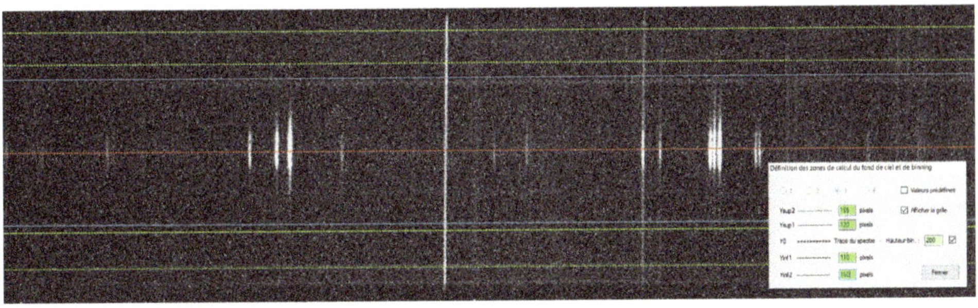

Fig. 4.4.3 — Binning and sky background zones in ISIS software.

The next tab "**3-Calibration**" (Figure 4.4.3) is used to define binning and background sky subtraction zones. The settings window is accessed by pressing the button "**Binning zone adjustement**".

The binning zone must include the entire trace of the nebula's emission lines. You can modify the image display thresholds so that edges are clearly visible (Figure 4.4.3).

Next, define the two subtraction zones for the background sky. These zones must include the interstellar or atmospheric lines that cross the entire spectrum, but not those specific to the nebula.

Then click on "**GO**" in the tab "**4-Go**". All the spectra from the nebula scan are processed, but ISIS keeps all the intermediate spectra @-1.fits, @-2.fits, etc. which will be of great importance later on.

All that remains is to reconstitute the image of the planetary nebula in the wavelength or wavelengths you have selected as a function of the emission lines present in the spectrum.

4.5 Image reconstruction

At this stage of treatment, you have all the data you need to produce an image in the wavelength of your choice among all the virtual filters whose width can be defined for each emission line. We therefore have access to the equivalent of a multitude of narrow-band filters, whose width can be much finer than commercially available imaging filters.

We see here the power of spectrography which enables an image to be obtained in any wavelength, whereas in conventional imaging, you would need as many filters as there are emission lines in the spectrum.

First you need to identify the different emission lines you can see on your spectrum. The graph in Figure 4.5.1 will help you spot them. The table in Figure 4.5.2 shows the elements and the wavelength of each at rest for the emission lines present in M57.

Depending on the planetary nebula being observed, the lines present may have different intensities or even be absent.

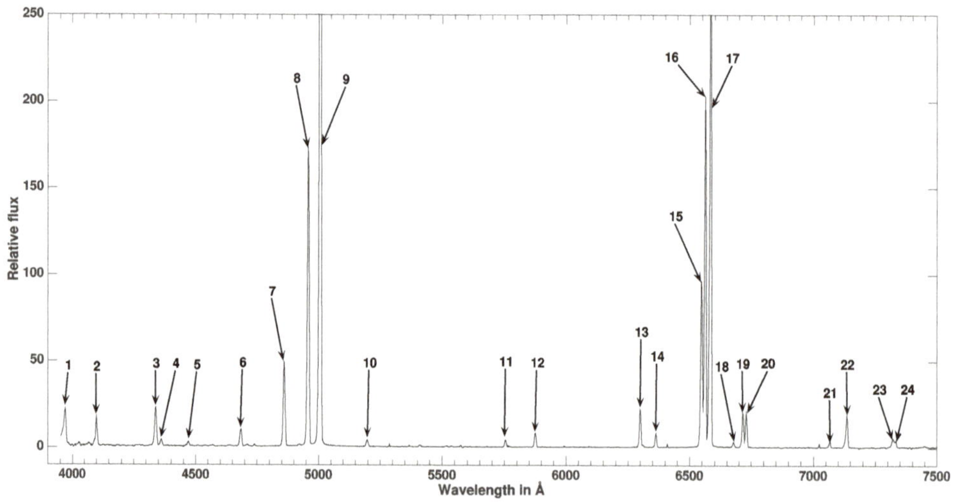

Fig. 4.5.1 — M57 spectrum. The most intense lines identified in the M57 spectrum. For greater visibility of the faintest lines, the intensity of [O III] 5 007 and [N II] 6 583 has been limited to 250. On the x-axis, wavelength in ångströms. On the y-axis, intensity relative to the continuum normalized to 1.

N°	Element	Value (Å)	N°	Element	Value (Å)
1	[Ne III]	3 967.41	13	[O I]	6 300.30
2	$H\delta$	4 101.73	14	[O I]	6 363.78
3	$H\gamma$	4 340.47	15	[N II]	6 548.03
4	[O III]	4 363.21	16	$H\alpha$	6 562.82
5	He I	4 471.48	17	[N II]	6 583.45
6	He II	4 685.68	18	He I	6 678.15
7	$H\beta$	4 861.33	19	[S II]	6 716.47
8	[O III]	4 958.92	20	[S II]	6 730.85
9	[O III]	5 006.84	21	He I	7 065.28
10	[N I]	5 197.90	22	[Ar III]	7 135.78
11	[N II]	5 754.64	23	[O II]	7 319.40
12	He I	5 875.67	24	[O II]	7 330.70

Fig. 4.5.2 — M57 emission line chart. The wavelength value corresponds to the "at rest" value. On other nebulae, we can see other lines of varying intensity.

Now, let us identify these lines on our images:

Open the 2D image of M57 spectrum, that calculated by ISIS in section 4.4. Push the visualization thresholds to see all the lines, as some of them are quite faint.

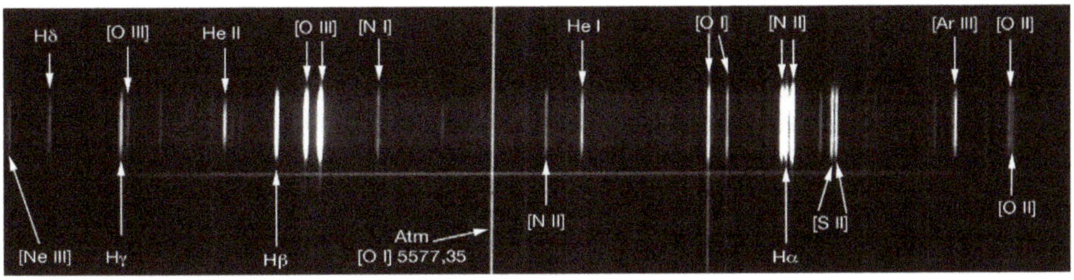

Fig. 4.5.3 — M57 2D spectrum. Identification of the main emission lines.

The line in the middle of the spectrum crossing the entire slit is the auroral line [O I] 5 577.35 Å (Figure 4.5.3). It is not part of the nebula, but forms in the Earth's upper atmosphere. However, this line is very useful: it can be used to check spectral calibration, as it is wavelength-stable. It will be removed from the useful signal during the background subtraction process.

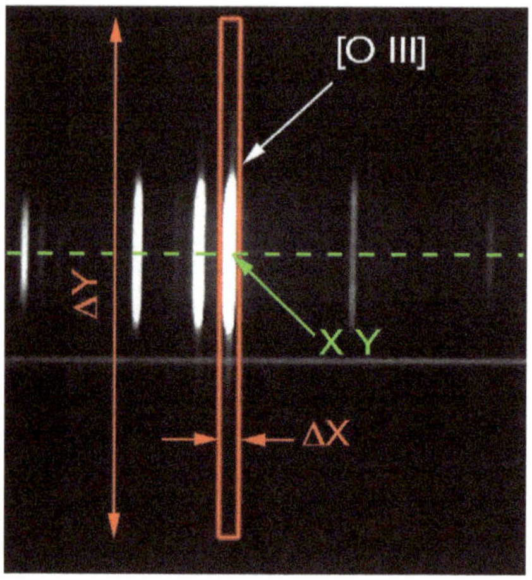

Fig. 4.5.4 — Defining a line's parameters. In green, the X, Y coordinates of a line center. In red, the line width in pixels ΔX and the line height ΔY.

To reconstitute the image at a given wavelength, we will use as an example the [O III] line at 5 006.84 Å (Figure 4.5.4). This procedure is the same for all other lines.

It is this oxygen line that is used on narrow band filters to produce SHO images. In the case of M57, it is also the most intense in the spectrum.

To reconstitute the image, use the command line part of the ISIS software by clicking on the "**Tools**" tab, then "**Command line**". But before you fill it in, you need to have measured the positions in pixels: of its center (X, Y), the width of the line in pixels (ΔX) and the height of the entire nebula (ΔY).

The function to be used here is **SCAN** whose syntax is as follows:

SCAN [SEQ] [Nb] [NAME] [X] [Y] [ΔX] [ΔY]

where:

[SEQ]: the generic name of the files calculated by ISIS for each slit position. In our case, the generic name is the @ symbol (This is the prefix of the name of the temporary 2D spectra generated by ISIS, located in the working directory: @-1.fits, @-2.fits...).

[Nb]: the total number of spectra in the planetary nebula. In this example, there are 80 spectra.

[NAME]: the name of the output image in .fits format. Feel free to name it anything you like, but for added convenience, you can name your file by starting with the name of the planetary nebula followed by the wavelength line. In this example, the file could be named: M57-OIII5007.

[X]: the X position of the selected line center, $X = 415$ in this example.

[Y]: the Y position of the selected line center, $Y = 591$ in this example.

[ΔX]: the line width in ångströms, this parameter affects the filter's width, its "passband", $\Delta X = 7$ in this example.

[ΔY]: the height of the nebula, it is important to get the full height of the nebula in the spectrograph slit, $\Delta Y = 235$ in this example. Think of the margins to leave a space between the edge of the image and the PN.

In this example, the command line becomes:

SCAN @ 80 M57-OIII5007 415 591 7 235

The result is an image file named M57-OIII5007.fits saved in the M57 working directory. This is the image of M57 at wavelength 5 007 Å.

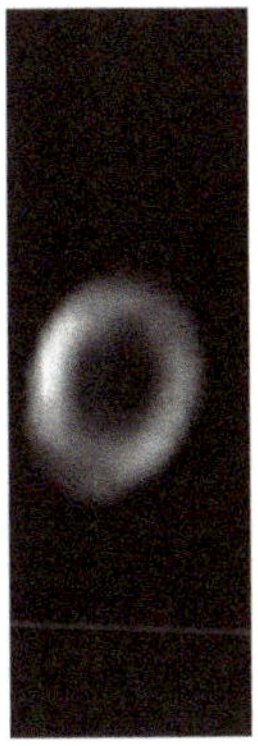

Fig. 4.5.5 — M57 image at λ = 5 007 Å.

Figure 4.5.5 shows the result of placing the 80 M57 spectra "side by side" shifted by 1 pixel for each position of the spectrograph slit on the line [O III] 5 007 Å.

Fig. 4.5.6 — Distorted image of M57 at λ = 5 007 Å. Scanning was performed in *Y* every 4 pixels.

Scaling the image:

If you have used an offset greater than one pixel in *Y* (horizontal slit), for instance 4 pixels (to save time during the acquisition process), the image size on this axis will be reduced by the same factor. The image will then be less defined in this direction and distorted (Figure 4.5.6). It will need to be rescaled along the *X* axis by multiplying the *X* image by a factor of 4. You can do this with a variety of image processing programs. But you can also simply request it from ISIS using the **SCALE** function with the following syntax:

SCALE [NameIN] [NameOUT] [*X*] [Y] 2

where:

[NameIN]: the name of the file to be resized.
[NameOUT]: the output file name (resized).
[*X*]: the horizontal scaling factor.
[*Y*]: the vertical scaling factor.

The last parameter "2" corresponds to the method used by ISIS to perform this scaling, in this case spline mode.

In the case of a 4-pixel offset in *Y* during acquisition, the image resizing function is:

SCALE M57-0III5007 M57-OIII-V2 4 1 2

We can also compare the image generated by ISIS with an image from the autoguiding sensor, by superimposing the 2 images and try, by successive iterations, to obtain the right scaling factors in X and Y.

The best results will be obtained by digitizing the planetary nebula in the finest possible detail (1-pixel shift at each scanning step) and by spending much more time acquiring single spectra (several exposures).

4.6 M57 at various wavelengths

By repeating this treatment for as many lines as you like, you will obtain a different image for each identified wavelength, which can then be combined.

Some very faint lines may not give an image with a good signal-to-noise ratio, only the most intense lines will give good results. You can also change the $[\Delta X]$ parameter, which modifies the line width to be taken into account. In a way, this reflects the width of the narrow-band filters used in imaging (3 nm, 5 nm and 7 nm).

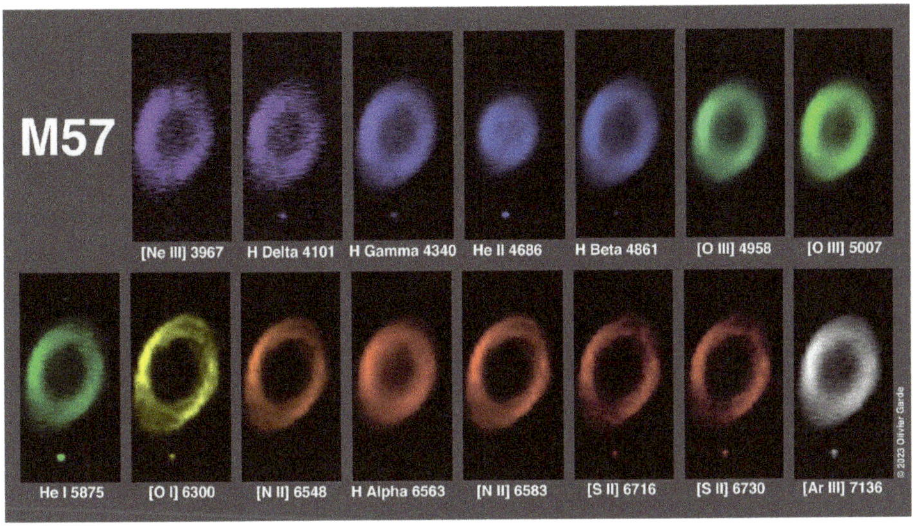

Fig. 4.6.1 — Multi-wavelength spectro-imaging of M57. The original images are, of course, in black and white, they have been coloured according to the wavelength of the observed element.

Depending on the characteristics of your spectrograph and camera, you can obtain images with bandwidths much narrower than 3 nm.

We can also combine 3 different wavelengths to reconstitute an image in pseudo-colour as is done in imaging with SHO filters. You can use conventional astronomical imaging software here, assigning each RGB layer of the final image, a wavelength of your choice. For example: red for [S II], green for $H\alpha$, blue for [O III] will produce an image of type "SHO".

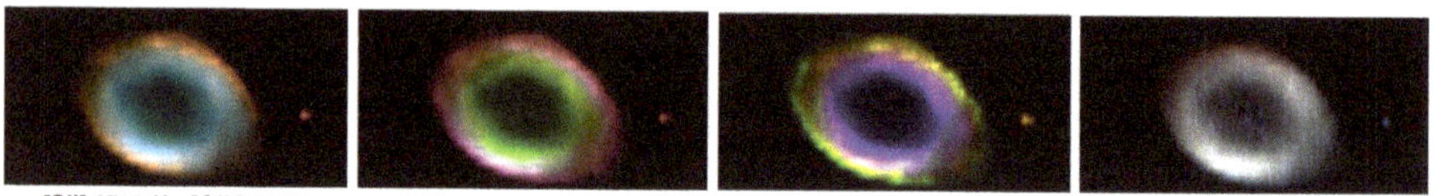

Fig. 4.6.2 — RGB colour image reconstruction. Some examples of different wavelength assemblies. On the left, the traditional SHO (sulfur, hydrogen, oxygen) assembly).

Depending on the wavelengths selected, we can reveal zones corresponding to a particular chemical element, represent the various lines of a single element ([O III] 5 007 Å, [O III] 4 958 Å, [O III] 4 363 Å) or hydrogen ($H\alpha, H\beta, H\gamma$) combined with 3 RGB layers. There is a huge number of combinations!

The resulting images are not necessarily very aesthetic but give an indication of the abundance of a chemical element in a specific area of the planetary nebula. For example, if we look at the He II 4686 A image in Figure 4.6.1, helium is only present in the center of the nebula and not on the ring.

4.7 Making science with planetary nebulae

In addition to being able to produce images as we have just seen, you can also perform some physical calculations on planetary nebulae. M57 will once again serve as an example. The calculations will be detailed for:

- The Balmer's decrement.
- The reddening constant.
- The electronic temperature.
- The electron density.

But we can also measure other parameters such as:

- The chemical abundances.
- The excitement class.
- The ionization structures.
- The expansion velocity (requires high-resolution spectra).

It is of course necessary to have well-calibrated relative flux spectra and to have correctly applied the instrumental response achieved at the same height as the planetary nebula (with a type O, B or A reference star).

It is difficult to obtain a single instrumental response, common to all spectra produced, because observations take a long time to make. Target height and airmass can vary significantly between the beginning and end of observations.

In order to limit this influence, is it preferable to select a nebula high enough in the sky during the acquisition period and start observing before meridian crossing, to limit the impact of the air mass on the spectral profile.

The air mass is the thickness of the atmosphere through which the light from the target passes. It is 1 at zenith, but rises to 2 at 30 degrees and then increases very rapidly.

This air thickness modifies the spectral profile, especially in the blue part of the spectrum, the spectrum becomes redder, which is why the calibration stars for instrumental response must be observed at the same height to compensate.

We have all noticed that the setting Sun or Moon is perceived as "red" because their light passes through a much thicker slice of the atmosphere, attenuating the blue part of the spectrum more strongly.

The simplified equation used to calculate the air mass M_a for a height h (in degrees) above the horizon is:

$$M_a = \frac{1}{\sin(h)}$$

It is also important to obtain spectra with a good signal-to-noise ratio, so as to be able to use the low-intensity lines essential for certain calculations without too many problems.

The calculations detailed later in this chapter are not necessarily very precise and scientific rigor could be improved, but the aim here is to show the methodology for obtaining a physical measurement generally comparable with the "pro" literature.

We can measure temperature or electron density from a spectrum and map them according to these values.

One last essential point for the calculations that will be detailed later in this chapter is to distinguish between 2 very different types of data:

• 2D image type .fits files such as acquisitions or those just calculated for image reconstruction.

• 1D spectra .fits files, (graphs) with wavelength on the x-axis and intensity on the y-axis.

In this example, the .fits file used is the average of all the spectra from the previous scan.

4.8 Reddening correction applied to a spectrum

When observing an object, the light it emits is modified by the presence of gas and dust in the line-of-sight between it and the observer. We call this phenomenon interstellar extinction, which "redden" the lines of a spectrum to a greater or lesser extent. It depends on the position of the observed object in the sky. It is the same principle as reddening due to air mass.

This phenomenon is observed, for example, when the sun sets at the horizon, where it is perceived as "red" by an observer. In this case, sunlight below the horizon passes through a much denser zone of the Earth's atmosphere. This "filters" the Sun's radiation by strongly attenuating the short wavelengths located in the blue part of the spectrum.

Short wavelengths are therefore attenuated to a greater or lesser extent depending on the distance from the target, its position in our galaxy and the height during the observation.

Let us take the example of the spectra made at the beginning of this chapter on M57. We begin by averaging or summing all the spectra to get a "global" view of the object and not a particular area of the planetary nebula.

The resulting 1D spectrum will be normalized to the value of 100 on the $H\beta$ line. This can be done simply by measuring the intensity of this line and then applying a coefficient to bring this value to 100. This multiplication coefficient, to be applied across the entire spectrum, is determined by the following simple equation:

$$Mc = \frac{100}{I_{(H\beta)}}$$

Where:

Mc: the multiplying coefficient factor to be applied to the spectrum.
$I_{(H\beta)}$: the intensity measured on the spectrum on the $H\beta$ line.

To measure the intensity of a line on a 1D spectrum, you can use the tools available in ISIS and Visual Spec:

• In Visual Spec: the function "**One gaussian fit**" from menu "**spectrometry**".

• In ISIS: the function "**FWHM**" of the tab "**5-Profile**".

Several spectrographic software packages can be used to perform simple arithmetic operations (addition, subtraction, multiplication, division). You can also do it in Python. Here is how it is done with ISIS software:

• Go to the menu "**Tools**", then "**Spectral**".
• Fill in the zone "**Multiply an profile by a constant**".

Figure 4.8.1 indicates the 3 fields to be filled in:

• Image: The name of the input source file.
• Coefficient: The multiplying coefficient factor (Mc) calculated above (2.016 in this example).
• Result: The name of the output target file.
• Click on "**GO**".

You now have a normalized spectral profile with an intensity value $I_{(H\beta)} = 100$ (the HB 100 file in this example).

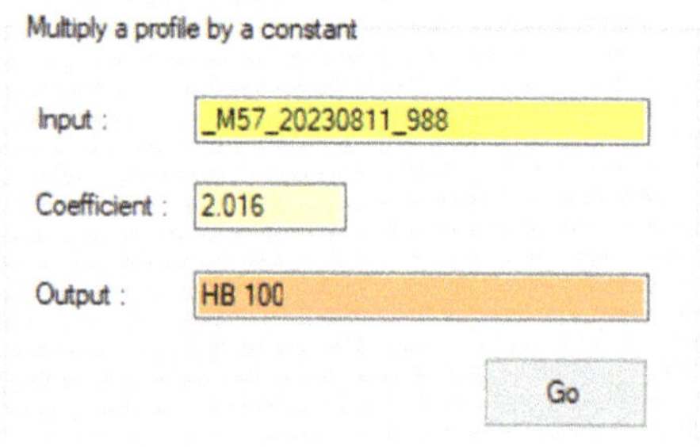

Fig. 4.8.1 — Window "Multiply a profile by a constant". ISIS software.

4.9 The Balmer decrement

It was the Swiss physicist and mathematician Jakob Balmer (1825-1898) who determined the mathematical relationship between the various wavelengths of the hydrogen atom in the visible range.

Balmer also found that the intensity of each line decreases from $H\alpha > H\beta > H\gamma$.

This decrease in intensity is called the Balmer decrement.

This decrement varies with temperature. In the case of planetary nebulae, which have an average temperature of 10 000 K, it takes the values shown in Figure 4.9.1, commonly accepted in various scientific publications.

Lines	λ (Å)	Ratio
$H\alpha$	6 562.82	2.85
$H\beta$	4 861.33	1
$H\gamma$	4 340.47	0.47
$H\delta$	4 101.73	0.26

Fig. 4.9.1 — Balmer decrement values table. For the case of a planetary nebula at 10 000 K.

In our practical case on M57, the next step is to measure the intensity of the $H\alpha$ line, make the division ratio with the $H\beta$ line and compare the observed value obtained with the theoretical value of 2.85, according to the following relationship:

$$\left(\frac{I_{(H\alpha)}}{I_{(H\beta)}}\right)_{Obs} > \left(\frac{I_{(H\alpha)}}{I_{(H\beta)}}\right)_{Theo} = 2.85$$

Some publications suggest a ratio of 2.86 between the intensities of $H\alpha$ and $H\beta$.

In real observations, this ratio is always greater than 2.85 due to interstellar extinction, but also to the extinction caused by the planetary nebula itself, depending on its morphology and density.

In the case of M57, extinction due to dust within the nebula itself is quite significant, increasing the reddening. This extinction is therefore not homogeneous and depends on the zone of the nebula to be analyzed. The following calculations were performed using the average of all spectra from the M57 scan, therefore the central and outer zones of the nebula mixed together.

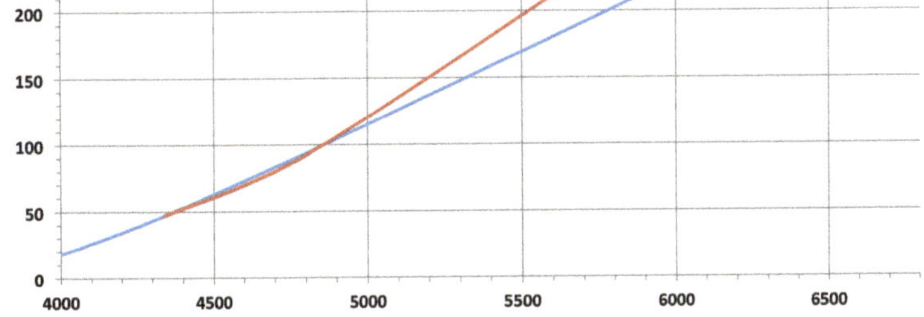

Fig. 4.9.2 — Theoretical and measured Balmer decrement. The red curve shows the measured decrement on the M57 spectrum, while the blue curve is the theoretical decrement. On the x-axis, wavelength in ångströms. On the y-axis, intensity of flux.

We will now correct the reddening of the M57 spectrum using the intensity ratio observed on the 2 hydrogen lines $H\alpha$ and $H\beta$ (with $H\beta$ normalized at 100). The extinction constant $c(H\beta)$ with respect to the $H\alpha$ line is given by the following equation (source: Astronomie Astrophysique 4th edition. Agnès Acker 2005):

$$c(H\beta) = (3.08)\ \log\ (I(H\alpha)_{obs}) - (7.55)$$

where:

$c(H\beta)$: the extinction constant with respect to the $H\alpha$ line.
$I(H\alpha)_{obs}$: the measured intensity of the $H\alpha$ line (with a spectrum normalized to $I(H\beta) = 100$).

It is possible to calculate the color excess $E(B\text{-}V)$ between the 2 photometric bands B (4 400 Å) and V (5 500 Å) with the relation:

$$E(B-V) = \frac{c(H\beta)}{(1.46)}$$

To deredden a planetary nebula spectrum, you can use ISIS software with the function "**Extinction**" available in the menu "**Misc**".

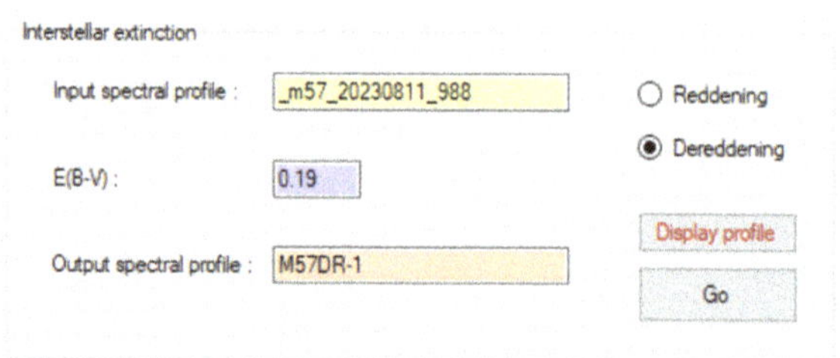

Fig. 4.9.3 — Window "Interstellar extinction" of ISIS software.

Figure 4.9.3 shows the various fields to be completed:

- **Input spectral profile**: the 1D file name of the spectrum to be corrected.
- $E(B\text{-}V)$: the excess color value calculated above.
- **Output spectral profile**: the name of the dereddening 1D file (here M57DR-1).

- Do not forget to tick the box "**Dereddening**".

We have just seen how to correct a reddening spectrum by applying the excess color coefficient to the entire spectrum. ($E(B\text{-}V)$). However, it is also possible to deredden a particular line with more precision with the following equations:

$$I_{(\lambda)} = I_{(Obs)} \times 10^{(c(H\beta)f(\lambda))}$$

For $f(\lambda)$, the equation was established by James B. Kaler (1976) on the basis of an observation curve made by Albert E. Whitford (1958):

$$f(\lambda) = 2.5634\lambda^2 - 4.8735\lambda + 1.7636$$

where:
$I_{(Obs)}$: the measured intensity value of the chosen line of wavelength λ.
$c(H\beta)$: the value of the extinction constant.
λ: the wavelength in μm.
$I_{(\lambda)}$: the dereddening intensity value of a λ line.

A practical example with the [O III] 5 007 Å line. We therefore measure its intensity on the 1D spectrum ($H\beta$ normalized to 100), here this value is $I_{obs} = 1\ 067$.

Numerical application of the above equations gives:

$$f(\lambda) = 2.5634 \times 0.5007^2 - 4.8735 \times 0.5007 + 1.7636 = -0.034$$

$$c(H\beta) = 0.32 \text{ (calculation based on the average of M57 spectra)}$$

$$I_{obs} = 1\ 067$$

$$I_{dr} = 1\ 067 \times 10^{(0.32x(-0.034))} = 1\ 041$$

The intensity value of the [O III] 5 007 Å dereddening line is 1 041 Å.

To determine the electron temperature or electron density, follow this procedure for each line required for the measurements ([N II], [S II] and [O III] lines). These calculations are described in the following pages.

4.10 Balmer decrement represented on an image

You can create an image of M57, showing the Balmer decrement measured in the nebula and therefore the influence of its density and interstellar extinction on the line of sight (see Figure 4.10.4).

To do this, you will use the images from the M57 scan for the two wavelengths

$H\alpha$ and $H\beta$. But in the images reconstructed from the scan, the instrumental response is not taken into account. It is therefore necessary to correct each one by applying a multiplying coefficient determined from the instrumental response curve used to calculate the 1D spectrum of the planetary nebula (with a nearby type 0, A or B reference star).

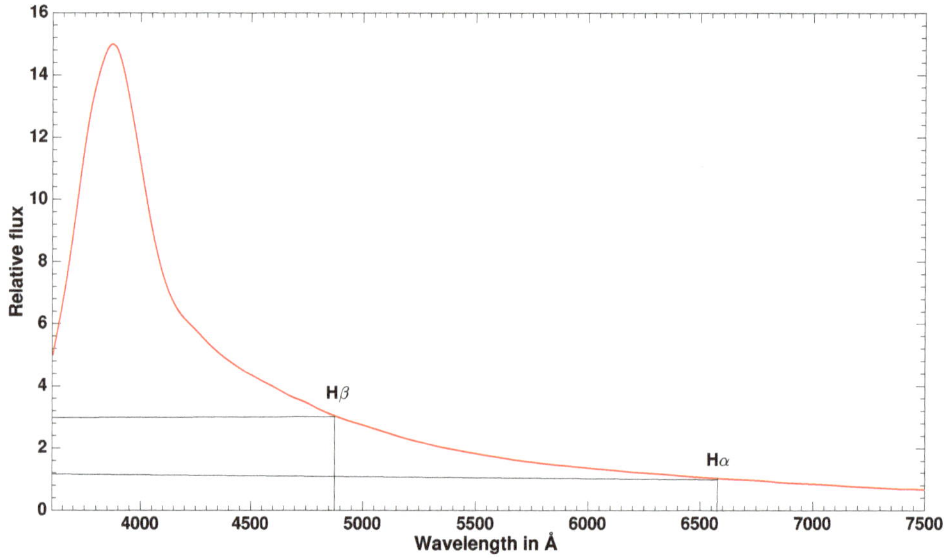

Fig. 4.10.1 — Using the instrument response curve. To determine the coefficient to be used to correct the intensity of an image as a function of its wavelength.The intensity ratio between the 2 emission lines will be used (intensity of $H\alpha$ and $H\beta$). On the *x*-axis, wavelength in ångströms. On the *y*-axis, intensity relative to the continuum normalized to 1 close to $H\alpha$.

You need to use on the response curve graph in Figure 4.10.1, the intensity value for $H\alpha$ and $H\beta$, then divide the 2 values to determine the multiplier coefficient to apply to the $H\alpha$ image:

$$M_c = \frac{I_{(H\beta)}}{I_{(H\alpha)}}$$

where:
$I_{(H\alpha)}$: the value of the intensity measured on the instrumental response at the $H\alpha$ position (6 563 Å).
$I_{(H\beta)}$: the value of the intensity measured on the instrumental response at the $H\beta$ position (4 861 Å).

M_c: the multiplying coefficient to be applied to the $H\alpha$ image.

Many astronomical image processing programs can perform mathematical operations to apply a coefficient to an image. Here is how to do it with ISIS (tab "**Tools**", then "**Images processing 1**").

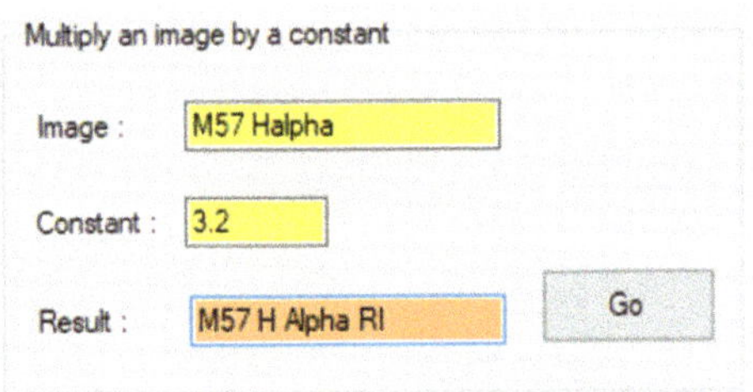

Fig. 4.10.2 — Window "Multiply an image by a constant". Of ISIS software.

Figure 4.10.2 shows how to multiplied an image by a constant with:

- Image: the .fit or .fits file name of the image to be multiply by the constant.
- Constant: the constant value (M_c).
- Result: the file name of the result of this operation in .fit or .fits format.

```
import numpy as np
import matplotlib.pyplot as plt
from astropy.io import fits
image_fileHa = "M57_H_Alpha.fits"
image_infoHa = fits.open(image_fileHa)
image_dataHa = image_infoHa[0].data
image_infoHa.close
image_fileHb = "M57_H_Beta.fits"
image_infoHb = fits.open(image_fileHb)
image_dataHb = image_infoHb[0].data
image_infoHb.close
offset=0.00001
image_calc = image_dataHa/(image_dataHb+offset)
plt.imshow(image_calc, cmap='nipy_spectral')
plt.clim(2,5)
plt.title('M57')
plt.colorbar(label=r'Rapport H$\alpha$/H$\beta$', location='bottom')
plt.xlabel('Pixel X')
plt.ylabel('Pixel Y')
plt.savefig('NP_Ha_Hb.pdf', dpi=400)
```

Fig. 4.10.3 — Python script for dividing 2 images.

In this case, we do not touch the $H\beta$ image, which is the reference image. We can then proceed to divide the images of $H\alpha$ by $H\beta$, which will show the Balmer decrement observed for this nebula (not corrected for reddening). Dividing two images can be done with most .fits image processing software and can also be done in Python with just a few lines of code.

The name of the $H\alpha$ image (corrected for instrumental response) is shown on line 4, that of $H\beta$ in line 8 (see Figure 4.10.3). The offset variable (on line 12) is used to ensure that no pixel value is set to zero in the denominator image of the division. This Python script generates an image showing the ratio of the division of the 2 spectra.

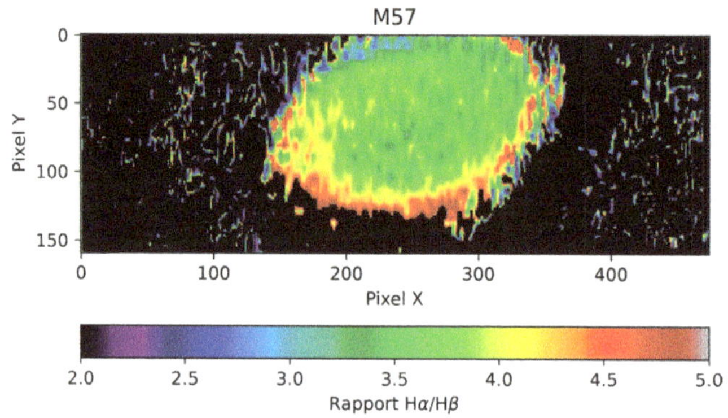

Fig. 4.10.4 — M57 Balmer decrement. The average ratio is close to 3.5 before spectrum dereddened. Denser areas experience more pronounced reddening.

The result can be improved, given the observation equipment used in this example (LISA spectrograph and a C14) and the short exposure time (600 seconds for each slit position on the nebula) but we're not far off the results obtained in some publications which show similar images with a ratio close to 3 to 4 on average (Performed with more powerful instruments such as spectro-imagers or integral field spectrograph; Lame, Pogge, 1994.)

We also note that the measured decrement is > 2.85.

4.11 Calculating the electronic temperature

The electronic temperature (noted T_e) is determined for a particular chemical element. In the case of a planetary nebula, this measurement is made on certain emission lines. They are called forbidden lines because they only appear under conditions of temperature and pressure not found on Earth. They are enclosed in square brackets: for example [O III].

There are several ways of measuring the electronic temperature of a planetary nebula using the [O III], [N II], [Ne III] or [S III] lines. Depending on the planetary nebula, one or other of the 4 series of emission lines can be used.

In most cases, measurements will be made with the [O III] and [N II] lines because the [Ne III] lines are located in the near UV range, hardly accessible to amateur spectrographs (lines at 3 869 Å, 3 968 Å, 3 343 Å). As for the [S III] lines, two of them are located in the near IR (9 532 Å, 9 069 Å) not easy either! But it is still possible with a spectrograph like the UVEX, which can cover both the near-UV and near-IR wavelengths.

In all cases, you need to measure the intensity of 3 lines, having first dereddened the spectrum as you described in section 4.8 above.

With this example on M57 and the LISA spectrograph, we will use the [O III] and [N II] lines. The dereddened values are shown in Figure 4.11.1 and were calculated on the average of all M57 scan spectra.

Line	λ (Å)	Intensity (I)
[O III]	5 007	1041
[O III]	4 959	339
[O III]	4 363	8
[N II]	6 583	449
[N II]	6 548	137
[N II]	5 755	6

Fig. 4.11.1 — Intensity values of the two series of lines. [O III] and [N II] calculated based on the average of the M57 spectra.

The problem that can be encountered when using [O III] lines is that the [O III] 4 363 Å line is quite weak and located very close to the $H\gamma$ line. Depending on spectrograph resolution, these two lines can be mixed (*blend*). We can also be embarrassed by the presence of the Hg 4 358 Å line if observed from a site with high light pollution.

If we use [O III] lines, the equation used to calculate the electronic temperature is as follows (Source: *Astrophysics of Gaseous Nebulae and Active Galactic Nuclei* - Donald E. Osterbrock and Gary J. Ferland - 2006):

$$R_{[\text{O III}]} = \frac{I(5\,007)+I(4\,959)}{I(4\,363)} = \frac{(7.90)\exp\left(3.29\cdot\frac{10^4}{T_e}\right)}{1+4.5\cdot 10^{-4}\left(\frac{N_e}{\sqrt{T_e}}\right)}$$

Which can be simplified in the case of a planetary nebula, by neglecting the denominator $(4.5)\ 10^{-4}\ N_e/\sqrt{T_e}$. The equation becomes:

$$T_e = \frac{3.29\cdot 10^4}{\ln\left(\frac{R_{[\text{O III}]}}{(7.90)}\right)}$$

Application with M57 numerical data:

$$R_{[\text{O III}]} = \frac{I(5\,007)+I(4\,959)}{I(4\,363)} = \frac{1\,041+339}{8} = 172.5$$

$$T_e[\text{O III}] = \frac{3.29\cdot 10^4}{\ln\left(\frac{172.5}{7.90}\right)} = \mathbf{10\,670\ K}$$

This result, derived from the average of 80 spectra of M57, is in line with the results found in some publications, as shown in the table in Figure 4.11.2 (average values on the M57 surface).

Garnett *et al.* (2001)	Lame - Pogge (1994)	Liu *et al.* (2004)
11 000 K	10 600 K	10 600 K

Fig. 4.11.2 — M57 electronic temperature. With [O III] lines, according to 3 different publications.

The [O III] 4 363 Å line is fairly weak, the result of the calculation can vary greatly depending on the precision with which the intensity of the line is measured. We can also use [N II] lines (6 583 Å, 6 548 Å and 5 755 Å). The [N II] 5 755 line is weak in intensity. We can use the equation:

$$R_{[\mathrm{N\,II}]} = \frac{I(6\,583) + I(6\,548)}{I(5\,755)} = \frac{(8.23)\exp\left((2.50)\dfrac{10^4}{T_e}\right)}{1 + (4.4 \cdot 10^{-3})\left(\dfrac{N_e}{\sqrt{T_e}}\right)}$$

or its simplified version by neglecting the denominator $4.4 \cdot 10^{-3}\ N_e/\sqrt{T_e}$ as in the equation with lines [O III]:

$$R_{[\mathrm{N\,II}]} = \frac{I(6\,583) + I(6\,548)}{I(5\,755)}$$

$$T_{e[\mathrm{N\,II}]} = \frac{2.5 \cdot 10^4}{\ln\left(\dfrac{R_{[\mathrm{N\,II}]}}{(8.23)}\right)}$$

By applying the numerical values of the intensities of the [N II] dereddened lines and measured on M57, we obtain:

$$R_{[\mathrm{N\,II}]} = \frac{I(6\,583) + I(6\,548)}{I(5\,755)} = \frac{449 + 137}{6} = 97.66$$

$$T_{e[\mathrm{N\,II}]} = \frac{(2.5) \cdot 10^4}{\ln\left(\dfrac{97.66}{(8.23)}\right)} = \mathbf{10\,106\ K}$$

This result can be compared with those found in the literature, as shown in the table in Figure 4.11.3 (values with M57 [N II] lines).

Liu *et al.* (2004)	Lame - Pogge (1994)	Guerrero *et al.* (1997)
10 200 K average	9 643 K average	10 800 K main ring

Fig. 4.11.3 — M57 electronic temperature. With [N II] lines, according to 3 different publications.

In general terms, electronic temperatures measured with [N II] lines are lower than with [O III] lines especially in the case of highly excited planetary nebulae, where the intensity of the He II 4 686 Å line is strong.

4.12 The electronic temperature map of a nebula

Just as you were able to generate a Balmer decrement map, you can also create an electronic temperature map of the nebula for a given chemical element.

The first step is to correct the various images required, taking into account the instrumental response, as we did for the Balmer decrement image, by calculating the multiplying coefficient for each wavelength value of the images required.

For measurements with [O III] lines, apply the multiplying coefficient to lines at 5 007 Å, 4 959 Å and 4 363 Å. In the case of electronic temperature with [N II] lines, the 6 583 Å, 6 548 Å and 5 755 Å lines will have to be adjusted.

Next, each image is dereddened, using a following multiplication factor, the ratio between the value of the intensity of reddening line with a dereddening line, for example in the case of [O III] 5 007 Å, we measured a reddening line at 1 067 and then calculated a dereddening line at 1 041, which gives a ratio of:

$$C_{[\text{O III}]_{5007}} = \frac{I_{dereddened}}{I_{obs}} = \frac{1\,041}{1\,067} = 0.975$$

We then multiply the image [O III] 5 007 Å by this coefficient $C_{[\text{O III}]_{5007}}$. Next, perform the calculation for all the other images required, then, to generate the electronic temperature map, use the Python script shown in Figure 4.12.1 which generates an image of the electronic temperature of M57 for the [N II] element with the equation described earlier in this chapter. It is adaptable.

The calculation is performed in this script with the [N II] lines, but it can also be modified for [O III] lines changing, of course, the equation in line 18 and the variables in line 17 (according to the wavelengths of the lines required).

For calculations with [N II] lines, line 4 uses the [N II] 6 583 Å image, line 8 the [N II] 6 548 Å image and line 12 the [N II] 5 755 Å image.

The offset variable in line 16 is used to avoid having pixel values equal to zero in the denominator of the division for image calculation.

```
import numpy as np
import matplotlib.pyplot as plt
from astropy.io import fits
image_file6583 = "m57_nii6583c.fits"
image_info6583 = fits.open(image_file6583)
image_data6583 = image_info6583[0].data
image_info6583.close
image_file6548 = "m57_nii6548c.fits"
image_info6548 = fits.open(image_file6548)
image_data6548 = image_info6548[0].data
image_info6548.close
image_file5755 = "m57_nii5755c.fits"
image_info5755 = fits.open(image_file5755)
image_data5755 = image_info5755[0].data
image_info5755.close
offset=0.000001
RNII = ((image_data6583+image_data6548)/(image_data5755+offset))+offset
image_calc=(25000/np.log(RNII/8.23))
plt.imshow(image_calc, cmap='nipy_spectral')
plt.clim(5000,25000)
plt.title('M57')
plt.colorbar(label=r'$T_e$ [N II]', location='bottom')
plt.xlabel('Pixel X')
plt.ylabel('Pixel Y')
plt.savefig('Te_M57.pdf', dpi=400)
```

Fig. 4.12.1 — Python script for electronic temperature calculation.

The result is rather noisy due to the lack of exposure time, particularly certain outliers (temperature > 20 000 K) that appear in the image division by the [N II] layer 5 755 Å, quite low in intensity and closer to noise.

Temperature measurements in areas with a good signal-to-noise ratio (main ring area) are close to those published by Lame and Pogge (1994), which presents a similar image with an average temperature of 9 643 K or the publication by Barker (1987) with an average temperature of 9 400 K.

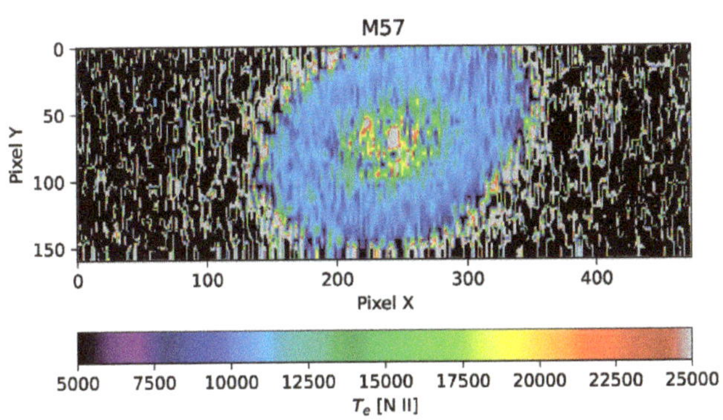

Fig. 4.12.2 — M57 electronic temperature map. Realized with the 3 lines [N II]: LISA spectrograph on a C14, ATIK 414EX camera in 1x1 binning. Temperature values are in degrees kelvin (K).

You can also see on the electronic temperature map in Figure 4.12.2 despite the noise in this part of the image, the central zone of the planetary nebula has a higher temperature than the main ring where the temperature is fairly homogeneous, contrary to what a simple "classical" image might suggest, with some areas more intense than others.

4.13 Calculating electron density

The second parameter that can be measured in planetary nebulae is electron density. This density is very low in this type of object, as the medium is very diluted, but very hot, with temperatures of around 10 000 K and more. To determine the electron density of a planetary nebula, we use a set of 2 lines with the same degree of ionization close to each other in wavelength.

The [O II] and [S II] lines are generally used for this type of calculation:

• For [O II], the 2 lines at 3 726 Å and 3 729 Å. These 2 lines are not very easy to obtain with amateur spectrographs because they are located in the near-UV range, but you can try to measure them with a UVEX spectrograph.

• For [S II], the 2 lines at 6 716 Å and 6 731 Å. If you use a spectrograph with low spectral resolution, we can not quite split these 2 lines. But with a LISA or UVEX spectrograph, with a nominal resolution of $R = 1\,000$, these 2 lines are well separated.

You can try to do the calculation from spectra made with a well-tuned Alpy 600 or a Star'Ex with an adapted grating. These two lines have the advantage of being located in the visible range, where amateur cameras perform well.

The equation used to determine electron density from the ratio of [S II] lines at 6 716 Å and 6 731 Å is given by the relation (source: *Astronomie, méthodes et calculs* by Agnès Acker and Carlos Jaschek, 1995):

$$R_{[\mathrm{SII}]} = \frac{I(6\,716)}{I(6\,731)} = (1.49)\left(\frac{1+(3.77)x}{1+(12.8)x}\right)$$

with:

$$x = 10^{-2}\, N_e T_e^{-1/2}$$

The electron density N_e is finally given by the relation:

$$N_e = 10^2 \sqrt{T_e} \left(\frac{R_{[\mathrm{SII}]} - (1.49)}{(5.62) - (12.8)\, R_{[\mathrm{SII}]}} \right)$$

where:
N_e: the electron density in cm^{-3}.
T_e: the nebula's electronic temperature in kelvin (K) (determined in the previous chapter).
$R_{[\mathrm{SII}]}$: the ratio of the intensities of the two lines $\frac{[SII]\ 6\,716}{[SII]\ 6\,731}$.

For the numerical application with the intensities measured on the two dereddening [S II] lines from M57:

$$I_{[6716]} = 28.95 \quad \text{et} \quad I_{[6731]} = 27.86$$

With a ratio of the 2 lines:

$$R_{[\mathrm{SII}]} = \frac{I(6\,716)}{I(6\,731)} = \frac{28.95}{27.86} = 1.039$$

This gives a density of:

$$N_{e[\mathrm{SII}]} = 100\sqrt{10\,106} \left(\frac{1.039 - 1.49}{5.62 - (12.8 \times 1.039)} \right) = \mathbf{590\ cm^{-3}}$$

Remember that the calculation was made using the average of all the spectra from the M57 scan and not just on one particular area of the nebula. Scientific publications indicate values of this order for the density in the main ring (table in Figure 4.13.1).

O'Dell *et al.* (2013)	Lame - Pogge (1994)	Wang *et al.* (2004)
550 cm^{-3}	700 cm^{-3}	525 cm^{-3}

Fig. 4.13.1 — Electron density of M57. According to 3 different publications and for the main ring.

We also note that the value of the electronic temperature (T_e), has a strong influence on the electron density result (N_e). For this calculation, we used the electronic temperature measured previously with the [N II] lines.

4.14 The density map of a nebula

As we did with the electronic temperature, we can reproduce an electron density map (noted N_e) of the nebula based on images [S II] 6 716 and 6 731 Å.

We will proceed in the same manner, namely readjusting the level of each image, taking into account the instrumental response, then unreddenned both images and adapt the Python script in Figure 4.12.1 with the correct equation (lines 17 and 18 of the script) to calculate the ratio of the two images in order to generate the electronic density map of the nebula.

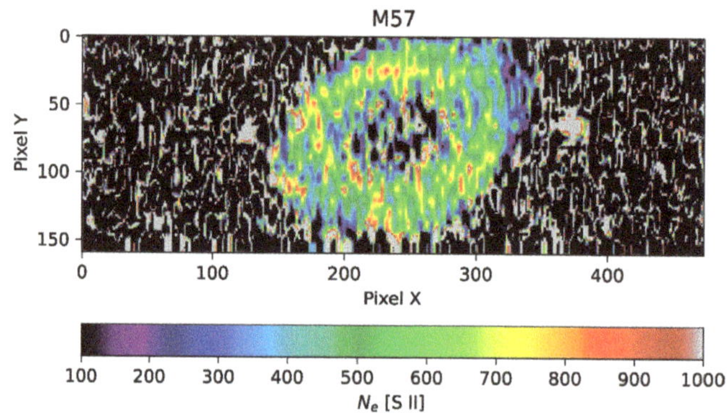

Fig. 4.14.1 — M57 electronic density map. Produced from the two lines [S II] with a LISA spectrograph, a C14 and an ATIK 414EX camera in 1x1 binning. Electron density values are in cm^{-3}.

As for the electronic temperature map, the result is rather noisy, but similar to what you will find in the literature.

You can repeat these measurements on different planetary nebulae and compare the results obtained. You can see the differences between each planetary nebula which are not all identical!

4.15 Bibliography

- Astronomie-Astrophysique, Introduction by Agnès Acker, Dunod, 2005. ISBN 2-10-005462-7.

- Astrophysics of Gaseous Nebulae and Active Galactic Nuclei by Donald E. Osterbrock and Gary J. Ferland, University Science Books, 2006. ISBN 1-891389-34-3.

- Astronomie: méthodes et calculs by Agnès Acker and Carlos Jaschek, Masson, 1995. ISBN 2-225-84689-8.

- Astronomical Spectrography for Amateurs: authors' collective, EDP Sciences, 2011. ISBN 978-2-7598-0629-4.

- Chemical abundances and kinematics of the ring nebula and its halos by M.A. Guerrero and A. Manchado. The Astrophysical Journal, 487: 328-345, 1997.

- Herschel images of NGC 6720: H_2 formation on dust grains by P.A.M. van Hoof et al., Astronomy and Astrophysics, 2010. DOI : 10.1051/0004-6361/201014590.

- The three-dimentional ionization structure and evolution of NGC 6720, the ring nebula by C.R. O'Dell, F. Sabbadin and W.J. Henney, The Astronomical Journal, 134: 1679-1692, 2007.

- Studies of NGC 6720 with calibrated HST/WFC3 emission-line filter images. II. Physical conditions by C.R. O'Dell, G. J. Ferland, W.J. Henney and M. Peimpert, The Astronomical Journal, 2013 (DOI: 10.1088/0004-6256/145/4/93).

- The law of interstellar reddening by A.E. Whitford, The Astronomical Journal, 1958.

- Proper plasma analysis practice (PPAP), an integrated procedure of extinction correction and plasma diagnostics: a demo with HST/WFC3 image set of NGC 6720 by Toshiya Ueta and Masaaki Otsuka, The Astronomical Society of the Pacific, 2021 (DOI: 10.1088/1538-3873/ac20ab).

- Electron density measurements in NGC 6720 by S.J. J. Czyzak, The Astrophysical Journal, 205: 798-801, 1976.

- IPHAS extinction distances to planetary nebulae by C. Giammanco and al., Astronomy and Astrophysics, 2010 (DOI: 10.1051/0004-6361/201014464).

- NGC 2392 spectroscopic analysis and François Teyssier:
www.astronomie-amateur.fr/feuilles/Spectroscopie/NGC2392.html

- spectra de-reddening by François Teyssier 2012:
www.astronomie-amateur.fr/Documents Spectro/Derougissement.pdf

- Spatially-resolved O II recombination line observations of the ring nebula, NGC 6720 by Donald R. Garnett et Harriet L. Dinerstein, The Astrophysical Journal, 2001.

- A catalog of relative emission line intensities observed in planetary and diffuse nebulae by James B. Kaler, The Astrophysical Journal, 31: 517-688, 1976.

- Electron temperatures in planetary nebulae by James B. Kaler, The Astrophysical Journal, 308: 322-336, 1986.

- Spectrocopic determination of electron temperature and electron density of planetary nebulae by B. Smarakoon et al., KDU Journal of Multidisciplinary Studies, Vol. 1, Issue 2, 2018.

- High-resolution spectra of the planetary nebula NGC 6803 by S.J. Lee and S. Hyung, Astronomy and Astrophysics, 2012 (DOI : 10.1051/0004-6361/201219263).

- Imaging spectrophotometry of the planetary nebula NGC 6720 by Nancy Joanne Lame and Richard W. Pogge, The Astronomical Journal, Vol. 108, Number 5, 1994.

- Numerical Python in Astronomy and Astrophysics by Wolfram Schmidt and Marcel Völschow, Springer, 2021. ISBN 978-3-030-70346-2.

5. Tracking Be stars

We should begin this chapter by asking the question: "What is a Be star?". The definition was popularised in 1987 by Georges W. Collins in an article entitled: "The Use of Terms and Definitions in the Study of be Stars" in which he repeats the terms of a 1981 publication (Jashek *et al.*): "Be stars are non-supergiant stars that have shown one or more Balmer lines in emission at least once during their lifetime".

Be stars are hot and blue (around 10 000 to 20 000 K) and represent about 20 % of all B-type stars. The letter "e" after the spectral type B, helps to distinguish between a B star and a Be star.

Be stars known as "classic" do not belong to luminosity classes Ia, Ib or II (so not to supergiants or luminous giants). They can only have a luminosity class of III, IV or V.

The first Be star was observed by Angelo Secchi (1818-1878) who mentioned, in a 1866 publication, that, unlike other stars of the spectral type B, γ Cas shows an $H\beta$ line in emission which he described at the time as "a curious peculiarity of this star". γ Cas is a fairly luminous Be star (magnitude $V = 2.47$) easy to identify in the sky, as it is at the center of the "W" of the constellation Cassiopeia. It shows very intense Balmer lines in emission. This emission is due to the presence of a rapidly rotating disk of matter around the star, but the physics of its formation are not yet well understood, hence the interest in being able to produce a large number of spectra of this type of star.

Here again, we see how powerful spectroscopy can be, providing more information than a simple photograph image. Only a spectrum shows that it's a Be star and not just a B-type star.

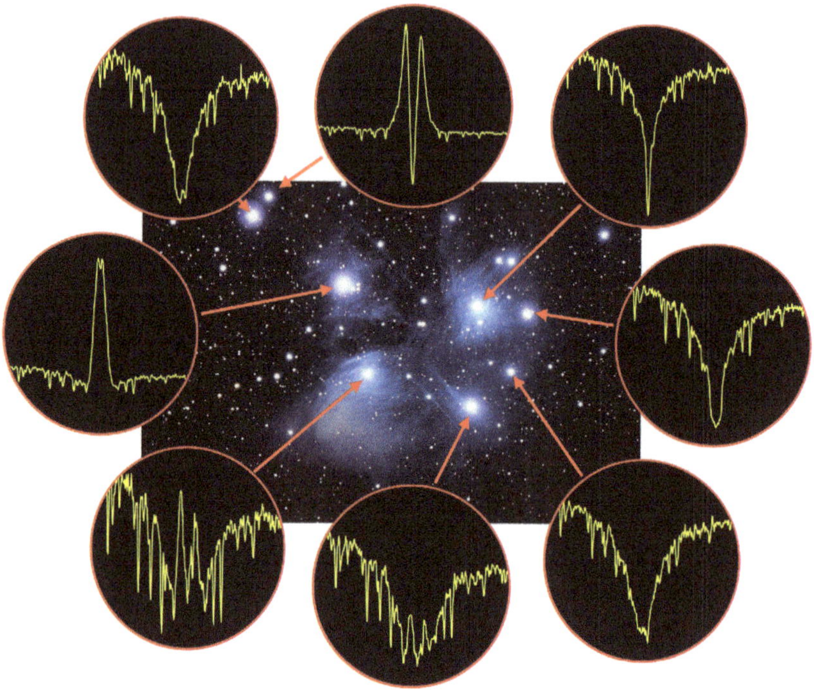

Fig. 5.0.1 — Spectra of M45 stars. High-resolution spectra (around the $H\alpha$ line) of the 8 brightest stars in the Pleiades.

Take, for example, the Pleiades open cluster (M45), which is mainly populated by young stars of spectral type B and luminosity classes III to V.

A simple photograph will reveal blue-coloured and therefore hot stars, but the image won't show you whether any of them are Be stars. Only spectroscopy will reveal those with Balmer lines in emission, as shown in Figure 5.0.1.

If we take the spectrum of the 8 brightest stars in this cluster, we can see the signatures of 4 stars with an $H\alpha$ line in emission: these are Be stars.

They are:

- Alcyone (25 Tau) B7III type
- Electra (17 Tau) B6III type
- Merope (23 Tau) B6IV type
- Pleione (28 Tau) B8V type

To study this type of star and its evolution, a pro/am collaborative program was set up by Coralie Neiner, astrophysicist at Paris Observatory. The beginnings of this collaboration were at evoked at the Oléron meetings in May 2003, with the idea of creating an affordable high-resolution spectrograph.

The LHIRES III was launched in 2005 in the form of a kit containing all the mechanical and optical components required for its operation. This was followed in 2007 by the creation of a spectral database: BeSS (Be Star Spectra) where anyone (professional or amateur) can upload their spectra.

Data from this database dedicated to Be stars have been used in several scientific publications and in some cases, contributors are listed as co-authors.

In this database there are, of course, the classic Be stars, but also 4 other types of star:

• The **Be-shell stars**: they are Be stars which also show deep, fine absorption lines of iron, titanium or chromium which arising from a shell of dense matter around the star on its equator. These lines can be temporary, and therefore change from a Be-shell type to a Be "simple" type.

• The **Herbig Ae/Be stars**: a class identified by Georges Herbig (1920-2013) and currently (2024) boasts 63 young stars that have not yet evolved into the main sequence. They are still surrounded by a disk of matter ejected by the star and have an intermediate mass of 1.5 to 8 $M_\odot$ (solar mass).

• The **B[e] stars**: are stars surrounded by a gaseous shell. Quite rare, there are 11 in the database (in 2024). To distinguish them from classic Be stars, the "e" is enclosed in square brackets. B[e] not Be!

• The **Be binary stars**: have one or more companions and include around 1/3 of all Be stars. The binary nature of these systems can create significant tidal effects on the Be star and affect its behaviour if the companion is close.

5.1 Observe the Be stars

Spectroscopic observations of a Be star are not difficult as the diversity of magnitudes means that a large number of targets are available, depending on the equipment used. Almost any type of spectrograph and any diameter of telescope can be used, but we can distinguish 3 different types of spectra to feed into the BeSS database.

5.1.1 Medium and high resolution

At medium and high spectral resolution (about $R > 9\,000$) it is to possible accurately record changes in the profile of a particular line. This can be obtained with spectrographs such as LHIRES III or Star'Ex HR (2 400 lines/mm grating), but also with Echelle-type spectrographs such as eShel or BACHES.

A single-order spectrographs such as LHIRES III or Star'Ex HR, must be set to a given wavelength, often the $H\alpha$ line, which is the most studied in high-resolution for this type of program, but other lines can also show interesting developments. The choice of line to study depends on the type of information required and may differ depending on the target and the type of study intended.

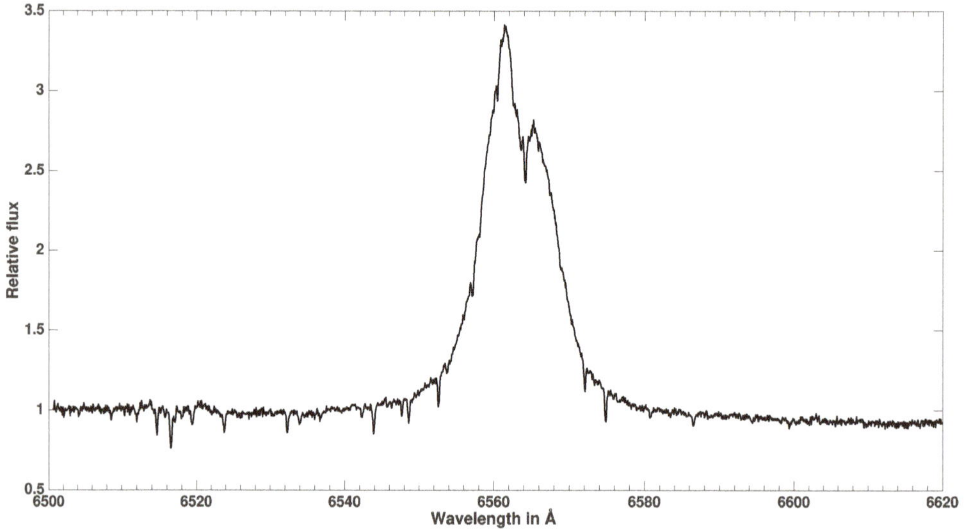

Fig. 5.1.1 — High-resolution spectrum of pi Aqr (magnitude V = 4.64). Centred on $H\alpha$. Realized with a $R = 30\ 000$ Echelle-type spectrograph on an RC telescope of 400mm in diameter (7 exposures of 900 seconds each). This spectrograph is not mounted on the telescope, but connected to it by a 50 µm-diameter optical fibre.

5.1.2 Echelle-type spectrographs

The advantage of a Echelle-type spectrograph is that it can cover the entire visible range in a single shot with medium/high resolution. This type of spectrograph is not necessarily within the financial reach of an amateur, but can be found in some astronomy clubs or mission observatories.

The only constraint with Echelle-type spectrographs is the magnitude limit of the target, which will be around 9-10 for a good signal-to-noise ratio with telescopes accessible to amateurs (typically up to 600 mm to 1 m in diameter, which can be found in mission observatories open to amateurs.

5.1.3 Low resolution

In low spectral resolution (between 500 and 3 000 approx.), which can be obtained with an Alpy 600, LISA, UVEX, Star'Ex LR (300 or 600 lines/mm grating). It is not possible to resolve detail within a single line. On the other hand, the entire visible spectrum can be captured, giving us a global view of the star. The great advantage of low resolution is that you can obtain spectra of stars with low luminosity and reach, depending on the optics used, up to magnitudes of 16-17 and even more fainter (with an Alpy 200 or SA 100). Low resolution complements high resolution by accessing targets unsuitable for high resolution with smaller diameter telescopes.

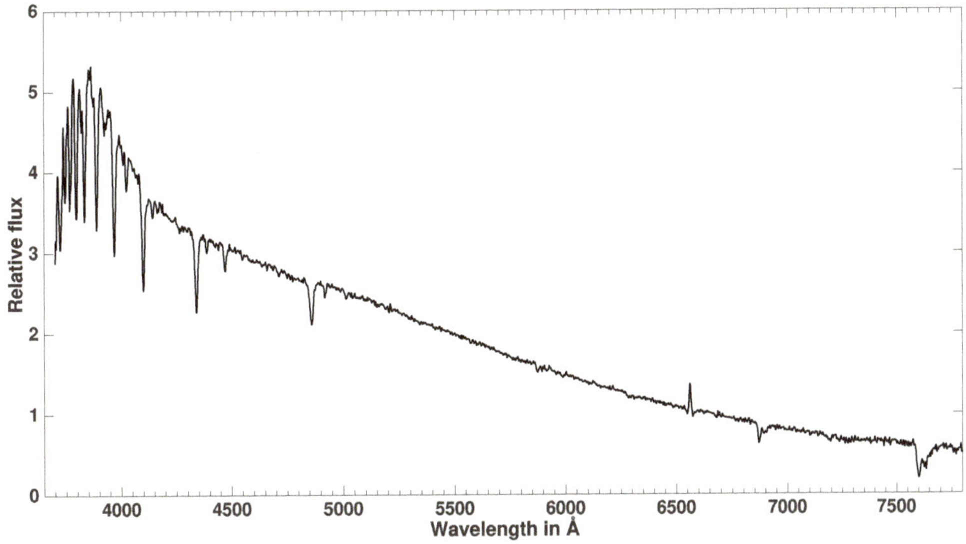

Fig. 5.1.2 — Low-resolution spectrum of BG Phe (magnitude V = 10.18). Captured with an Alpy 600 spectrograph equipped with a 23 μm slit on an RC 305 mm telescope (4 exposures of 600 seconds with an ATIK 414ex CCD Camera).

The magnitude of the target should therefore be chosen according to the resolution of the spectrograph and the diameter of the telescope. For example, there is no point in making a low-resolution γ Cas spectrum, because this target is already very well covered in high resolution, and low resolution does not provide any additional information. On the other hand, low resolution will be very useful for Be stars that have not yet been observed in the BeSS database due to their low luminosity. By the end of 2024, there will still be a thousand Be stars without a spectrum in the database. It is also used to monitor these stars to warn of a change of state: *outburst* or quiet phase.

5.2 Selecting targets for observation: the BeSS database

You need to determine which targets should be given priority, or which require special attention. To date (in 2024), the BeSS database contains more than 2 400 Be-type and related stars. A special tool, ArasBeAm, has been set up to indicate to the observer which stars require a spectrum for the BeSS program. This tool can be consulted online at the following address:

http://arasbeam.free.fr/?lang=en

It is updated daily with the latest spectra validated in the BeSS database. Here is how to use it:

- Once on the home page, click on "**Be list**" in the left-hand interface column.

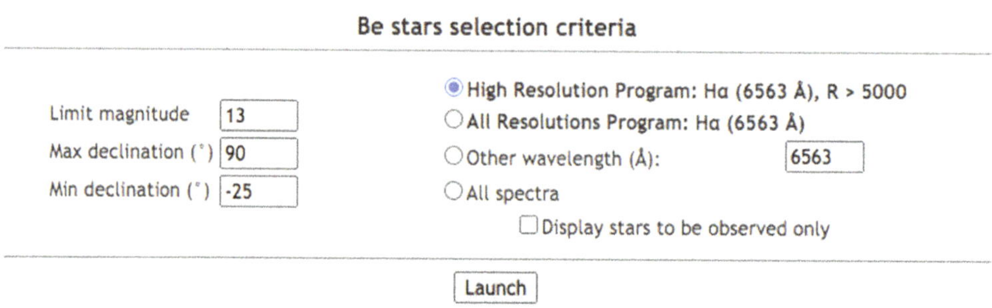

Fig. 5.2.1 — BeSS database selection criteria page.

- You can select a number of parameters to refine your search such as target declination limits, maximum magnitude and required resolution (see Figure 5.2.1).

Star	HD #	RA	DEC	Magn.	Sp. type	Tot. nb	1 year	2 months	Last	Obs Period
+ -	+ -	+ -	+ -	+ -	+ -	+ -	+ -	+ -	+ -	+ -
BD+62 2346	0	+00 01 24.7	+63 30 15.7	9.73	B0Ve	5	2	0	2025-01-15 20:46:04	365
HD 224905	224905	+00 01 38.6	+60 26 59.7	8.47	B1Vne	17	2	0	2024-12-25 21:12:23	365
HD 225095	225095	+00 03 27.1	+55 33 03.2	7.95	B2IVne	61	23	0	2025-02-22 18:49:50	INTENSIVE
2 Cet	225132	+00 03 44.4	-17 20 09.6	4.54	B9IVne	43	3	0	2024-08-31 07:29:47	365
10 Cas	144	+00 06 26.5	+64 11 46.2	5.57	B9IIIe	63	2	0	2024-11-22 20:54:37	365
V787 Cas	13590	+02 15 13.0	+64 01 28.0	7.90	B2IIIe	47	3	0	2025-01-12 03:28:40	120
BD+62 1	0	+00 07 33.2	+63 04 49.2	10.46	B2IVpe	1	0	0	2020-11-07 01:38:10	365
TYC 4306-1125-1	0	+00 08 03.0	+73 32 35.6	9.66	B8Ve	0	0	0		365
IP Per	278937	+03 40 47.0	+32 31 53.7	10.36	A6e	2	0	0	2019-09-08 03:56:30	365
HD 22780	22780	+03 41 07.9	+37 34 48.7	5.54	B7Vne	134	9	0	2025-02-04 17:14:20	90
13 Tau	23016	+03 42 18.9	+19 42 00.9	5.68	B9Vne	57	3	0	2025-01-13 17:26:34	365
ELECTRA	23302	+03 44 52.5	+24 06 48.0	3.71	B6IIIe	141	14	0	2025-02-22 22:26:54	1
BD+61 39	0	+00 20 17.4	+62 27 49.9	8.85	B0.5IVe	15	0	0	2023-09-08 08:51:39	365
BD+63 48	0	+00 26 40.1	+64 25 55.4	9.26	B1IIInne	1	0	0	2020-09-18 00:02:41	365
BD+55 81	0	+00 28 12.0	+56 23 47.5	10.02	B1.5Vnne	1	0	0	2020-11-17 22:48:18	365
BD+62 89	0	+00 29 49.2	+63 04 04.0	10.25	B0Vpe	1	0	0	2020-09-17 21:06:11	365

Fig. 5.2.2 — List of stars in the BeSS database. Displayed according to criteria that can be set from Figure 5.2.1.

The result of your query (Figure 5.2.2) will be color-coded to indicate whether or not a spectrum is urgently needed:

- In **green**: no particular needs.
- In **yellow**: a spectrum becomes necessary as the last one produced approaches the observation cut-off date determined from the period indicated in the last column.
- In **red**: a spectrum is required as the last one exceeds the observation period. If this box does not indicate a date, there is not a spectrum in the database yet, so you will be the first to produce one.
- In **violet/magenta**: we need to make intensive observations of this star. At least one spectrum per day is required. For this, high resolution is required, as we want to observe rapid changes in a particular line.

5.3 Submit your first spectrum to BeSS database

Once you have processed your first Be star spectrum, you can submit it in the BeSS database. It must be calibrated for wavelength and corrected for instrumental response. However, it is advisable not to correct the heliocentric or barycentric velocity, as this is a calculation that can be made later by the professional who will use it for his study. All the necessary information (date and time of observation, geographical coordinates of the observation site, target coordinates) can be found in the .fits header or in the observation sheets to be completed in at the time of submission.

Before submitting your spectrum to the database, you can check it with the Visual Spec software, which has a function for verifying the accuracy of your wavelength calibration, but also to compare it with one already validated in the BeSS database. To perform this check, open the spectral profile in Visual Spec, then click on the BeSS logo icon in the bottom right-hand corner of the window (of course, your PC must be connected to the Internet).

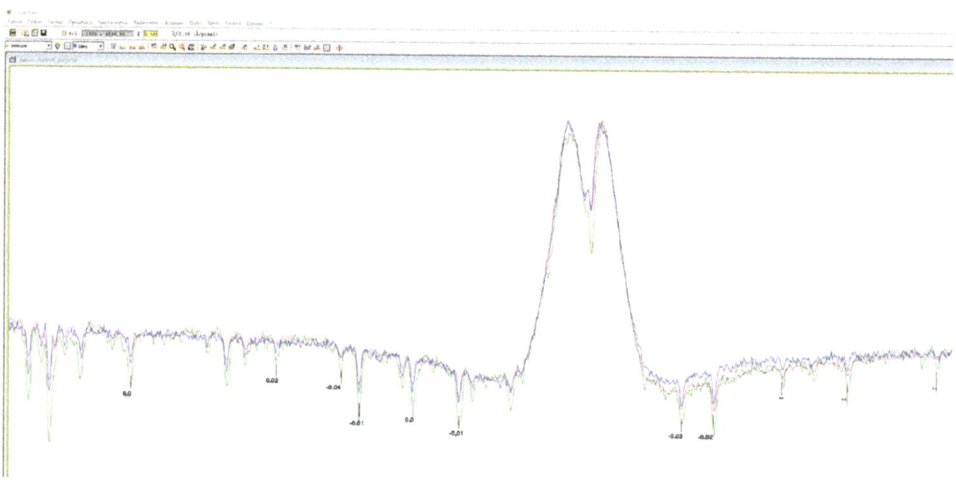

Fig. 5.3.1 — Verification of wavelength calibration. With Visual Spec software for high-resolution spectra.

Your spectrum appears in blue and is superimposed onto other validated spectra from the BeSS database here in green and magenta. (see Figure 5.3.1).

• In **high resolution**, the numbers shown on the graph correspond to the wavelength shift (measured - expected) in ångström (Å) of certain atmospheric lines around the $H\alpha$ line. If some values are greater than 0.1 Å, your spectrum calibration could be improved. If the values are less than 0.05 Å or even less, the calibration is accurate. This image can also be used to check whether the instrumental response is correct and see if there have been any significant changes in the $H\alpha$ profile.

• In **low resolution**, verification can be carried out using the same software and methodology. Wavelength calibration is then checked against the theoretical position of the hydrogen Balmer lines (green vertical lines on Figure 5.3.2).

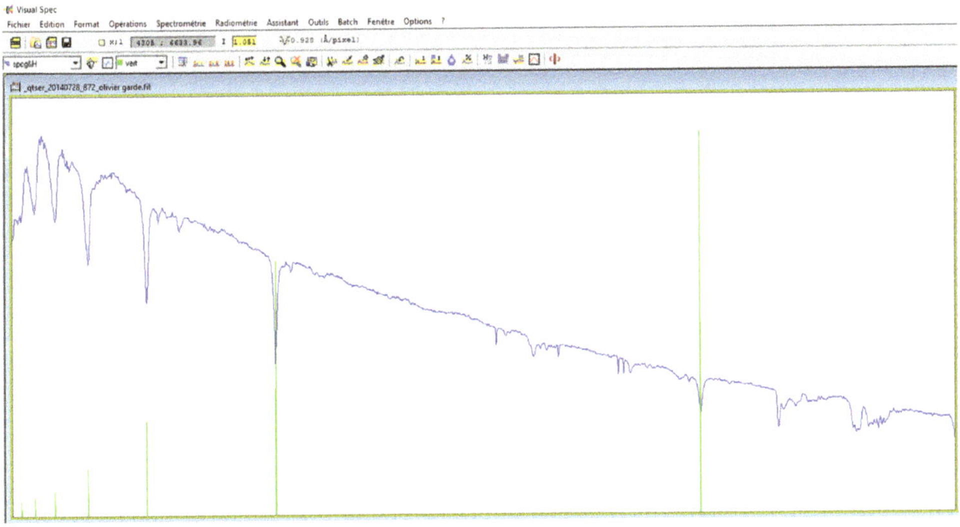

Fig. 5.3.2 — Verification of wavelength calibration. With Visual Spec software for low-resolution spectra.

Note that in low resolution, you will not have the same level of precision, but at the very least the lines should be at their nominal values.

These quick checks ensure that your spectrum is in compliance. From there, you can submit it to the BeSS database. To do so, visit the BeSS database website:

http://basebe.obspm.fr/basebe/

In the left-hand column of the interface, click on "**Registrations**". An initial window appears, offering you the opportunity to have your spectrum checked before submission (e.g.: compliance of fits header with BeSS standard). Then click at bottom left on "**Continue**". You must fill out 3 forms to submit your spectrum:

• The "**Observer**" form: The name of the observer alias must be the same as the one entered in the fits header of your spectrum. By validating this form, you will have access to a new page allowing you to upload your spectrum in .fit or .fits format.

• The "**Instrument**" form which describes the optics, spectrograph and spectral camera used. Please indicate the diameter of your optic in meters (e.g.: 0.2 m for a 200 mm telescope). The name given to this instrument sheet must be scrupulously identical to what is indicated in the fits header of your spectrum. Even if some fields on this form are not mandatory, it is important to provide as much information as possible.

• The "**Observation site**" form with longitude, latitude and altitude coordinates. Longitude is positive east of the Greenwich meridian and negative west of it. Latitude is positive for the northern hemisphere and negative for the southern. The name given here must be identical to what is indicated in the fits header of your spectrum. Do not duplicate: if you are observing from a known site (e.g. the Haute-Provence observatory or the Astroqueyras observatory at Saint-Véran in france), a "site" record is already present.

In all cases, check and use the existing data forms (observer, instrument, site) if the configuration is the same from one spectrum to the next. This is often the case with mission observatories, or at home. This will avoid duplicates and saves you time.

Once your first spectrum has been submitted, a team of validators will check it, and if necessary advise you. If it conforms, you'll receive your login and a password to submit subsequent spectra more easily.

5.4 The various profiles of Be stars

A Be star can have various types of profile: single peak, double peak, a peak in an absorption profile or no peak at all. To what do these various scenarios physically correspond?

The spectral profile of a line gives information about the inclination of the disk surrounding the star with respect to the line of sight. There are several possible cases (see Figure 5.4.1):

Case 1 corresponds to the spectrum of a classic B-type star, or a Be star outside outburst, which shows no emission. There is therefore no disc rotating around the star (not in outburst). In practice, it can be seen that the bottom of the absorption line has some irregularities and is not as "smooth" as the classical profile of a hydrogen line.

Cases 2, 3 and 4 show some of the spectral profiles that can be observed depending on the apparent inclination of the disk. Other intermediate profiles also exist. Emission amplitudes can be very different from one star to another. The inclination (i) is the angle of the rotation axis with respect to the line of sight (see Figure 8.0.1).

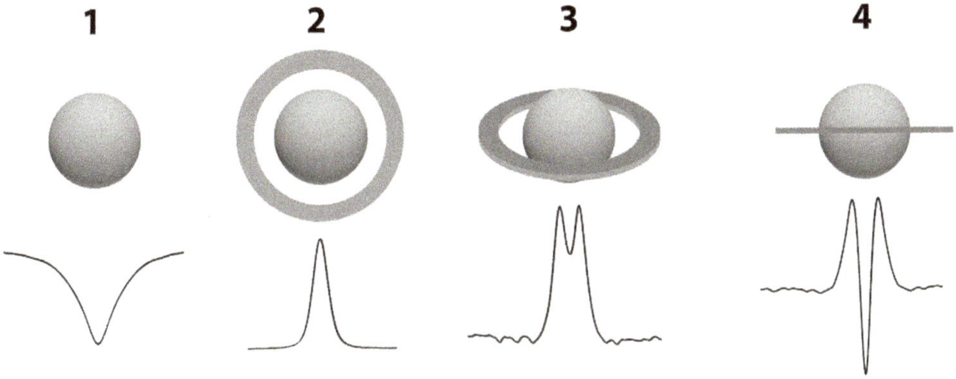

Fig. 5.4.1 — Some spectral profiles of Be stars. $H\alpha$ line spectra made with an eShel spectrograph $R = 11\ 000$.

The existence of a double peak corresponds to the Doppler-Fizeau effect. A shift between the part of the disc approaching the observer (blue shift) and the part moving away (red shift). We can, from this wavelength shift, measure disc rotation velocity and thus calculate the quantity $V\sin(i)$ corresponding to the measured velocity weighted by the inclination (i). Please note: this velocity is not the absolute velocity rotation (V_r) of the disk, as this requires knowledge of the inclination of the star's axis of rotation.

$$V_{Mes} = V_r \sin(i)$$

where:

V_{Mes}: the velocity measured on the spectrum.

V_r: the absolute velocity rotation.

i: the inclination of rotation axis in relation to the line of sight.

These 2 peaks are named V and R respectively:

- V for "Violet": the blue/violet-shifted peak of the spectrum.
- R for "Red": the red-shifted peak of the spectrum.

With this information, we can return to the last 3 cases:

Case n° 2: only one peak is visible, due to the emission from the disc seen from above, no part approaches or moves away on the line of sight. This is also the case when studying galaxies when viewed face on. The inclination is then close to 0 degrees, and as sin (0°) = 0, the real rotation velocity can take on any value! On the other hand, the width of the line gives information on density and the interactions between the hydrogen atoms in the disk.

Case n° 3: the disc is seen from an "oblique" angle (i.e. 45 degrees). The profile is intermediate between cases 2 and 4, the central absorption section deepens as the incline increases.

Case n° 4: spectrum of the 28 Tau star (Pleione) of the Pleiades cluster. The disk of this star can be seen almost edge-on. Two emission peaks and one absorption peak in the center. Yet another piece of information not available from a simple colour photograph!

5.5 Detecting an outburst

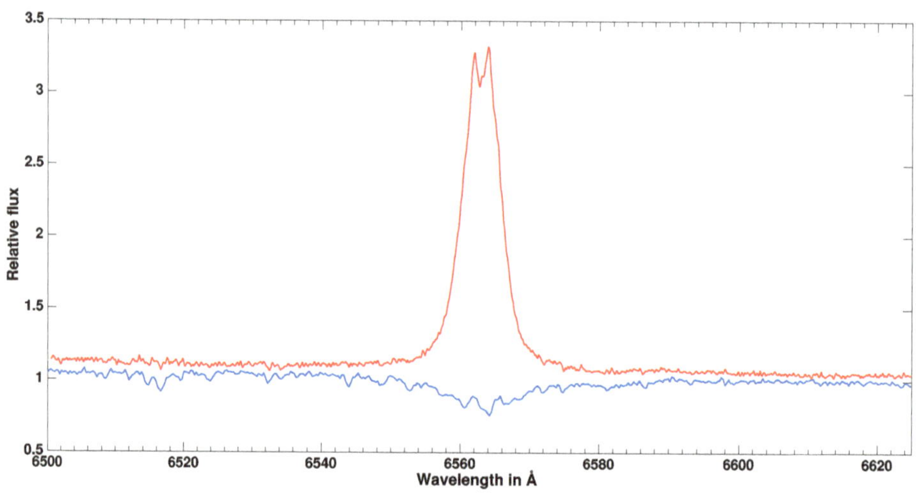

Fig. 5.5.1 — HP CMa outburst. The blue spectrum was created by Christian Buil in 2009, the spectrum in red by the 2SPOT team in 2023. Christian's spectrum was produced using a LHIRES III, and 2SPOT's spectrum using an eShel spectrograph.

Be stars change over time, and all are considered to be variable stars. Some are quite stable and show only slight modifications, others, on the contrary, show a rapid evolution of their spectral profile, even the disappearance of an emission or an unexpected appearance, sometimes over fairly short periods. When the spectrum changes from an absorption profile to an emission profile, we say that the star has gone into outburst. This kind of phenomenon is interesting to detect and attempt to find a recurring period between 2 appearances and disappearances.

Some Be stars have relatively regular and frequent outbursts, such as V442 And, QR Vul or OT Gem. Others are more erratic, sometimes with very long periods. To be informed of news from the Be stars, Valérie Desnoux has been writing a monthly newsletter since 2012 (*BeSS Monthly reports*). The archives are available here:

http://valerie.desnoux.free.fr/BeSS/index.html

The newsletter is also distributed monthly on spectroscopic mailing lists such as spectro-l, astronomical spectroscopy or Solex-project (through the groups.io platform) and also sent to professionals working on Be stars.

In the same way, outbursts can be detected at high resolution, providing maximum detail on the state of the star. We can sometimes detect the beginnings of this by observing the subtle changes within a profile. They can also be spotted on low-resolution spectra with a global view of the entire visible range but of course without so much detail on a particular line. This low-resolution observation can still be used to record the state of a Be star at a given moment.

5.6 Long-term measurement and monitoring

It is interesting to follow Be stars over several months, even years or decades to observe the temporal evolution of various parameters on Balmer lines, but also on other lines such as He I 6 678 Å, He I 5 876 Å, C II 6 578 Å, C II 6 583 Å. For some Be stars, this evolution can be very significant but sometimes very minor, up to several years, which is why it is a good idea to keep an eye on them regularly. Certain targets can be tracked over a specific period of time during a joint measurement campaign between space-based and ground-based observations. For example, with the satellite: TESS (*Transiting Exoplanet Survey Satellite*) in service since 2018... and still active at the time of writing this book (in 2024).

Requests for observations on a particular star can also be launched by astrophysicists. This was the case, for example, with V442 And (HD 6226) which allowed, thanks to the contributions of many amateur astronomers, the publishing of the article shown in Figure 5.6.1 in which all contributors were cited as co-authors.

Outbursts and stellar properties of the classical Be star HD 6226

Noel D. Richardson,[1*] Olivier Thizy,[2] Jon E. Bjorkman,[3] Alex Carciofi,[4]
Amanda C. Rubio,[4] Joshua D. Thomas,[5] Karen S. Bjorkman,[3] Jonathan Labadie-Bartz,[4]
Matheus Genaro,[4] John P. Wisniewski,[6] Luqian Wang,[7] Douglas R. Gies,[7]
S. Drew Chojnowski,[8] Andrea Daly,[1] Thompson Edwards,[5] Carlie Fowler,[5]
Allison D. Gullingsrud,[3] Nolan Habel,[3] David J. James,[9] Emily Kehoe,[5]
Heidi Kuchta,[3] Alexis Lane,[1] Anatoly Miroshnichenko[10,11,12] Ashish Mishra,[3]
Herbert Pablo,[13] Maurice Peploski,[5] Joshua Pepper,[14] Joseph E. Rodriguez,[15]
Robert J. Siverd,[16] Keivan G. Stassun,[17] Daniel J. Stevens,[18,19,20] Jesica L. Trucks,[21] James Windsor,[22]
Mackenna Wood,[5] Étienne Bertrand,[2] Jean-Jacques Broussat,[2] Erik Bryssinck,[2]
Christian Buil,[2] Stéphane Charbonnel,[2] Arnold de Bruin,[2] Joe Daglen,[2] Valerie Desnoux,[2]
James Dull,[23] Olivier Garde,[2] Keith Graham,[2] Kevin Gurney,[2] Alun Halsey,[2]
Patrik Fosanelli,[2] Joan Guarro Fló,[2] Franck Houpert,[2] Foster James,[2] Christian Kreider,[2]
Robin Leadbeater,[2] Tim Lester,[2] Dong Li,[2] Alain Maetz,[2] Albert Stiewing,[2]
Peter Somogyi,[2] Jean-Noël Terry,[2] Stéphane Ubaud,[2] and Ulrich Waldschlaeger[2]

Fig. 5.6.1 — The authors of the publication on V442 And.

This publication is available here: https://arxiv.org/pdf/2109.11026.pdf

Often these measurements can only be made on medium- and high-resolution spectral profiles, therefore typically performed with spectrographs such as the Star'Ex HR, the LHIRES III or an Echelle-type spectrograph (eShel or BACHES, for example). Measurements are carried out mainly on the $H\alpha$ line, as it is very bright and therefore easy to obtain a good signal-to-noise ratio.

Hydrogen is a major constituent of stars, and for the amateur, it is easy to calibrate the wavelength range around $H\alpha$ with precision, on the one hand, using a simple Neon lamp which has 3-4 emission lines in this zone but also by using atmospheric water lines (H_2O) present on the spectrum around the $H\alpha$ line.

However, it is not easy to calibrate a spectral profile centred on the $H\beta$ line which is located near the maximum sensitivity of CMOS spectral cameras (around 500 nm). This is due to the lack of reliable and inexpensive spectral lamps emitting lines around $H\beta$ and the absence of atmospheric lines in this zone. On the other hand, an eShel or Baches spectrograph are often equipped with a thorium/argon lamp, producing a large number of lines. In this instance calibration of medium and high-resolution spectra presents no issue in the entire visible range.

Many parameters can be measured on a line, such as rotation velocity, intensity of spectral peaks, and their evolution over time.

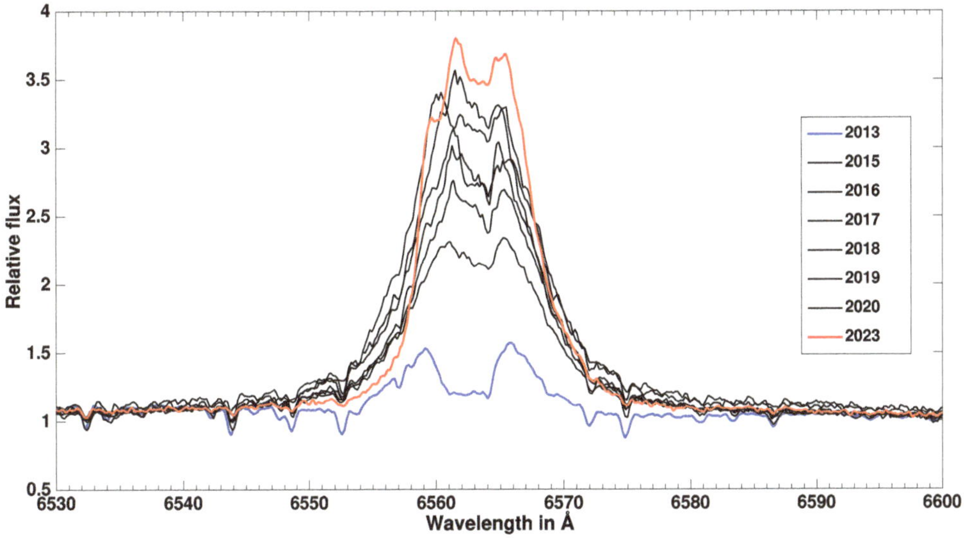

Fig. 5.6.2 — Pi Aqr spectrum evolution over 10 years. Series of spectra centred on the $H\alpha$ line, uncorrected for heliocentric velocity and using the same eShel spectrograph: resolution $R = 11\ 000$. Each spectrum is normalized to 1 on the continuum near the $H\alpha$ line.

The series of spectra shown in Figure 5.6.2. shows the evolution of the Be star pi Aqr over 10 years. We can see that the overall profile changes in intensity and the V and R peaks show intensity variations in relation to each other, but also that the wavelength distance between them also varies over time.

We will now describe two different measurements that can be carried out with time tracking.

- Measuring the V/R ratio of a spectral line and its variations.

- Equivalent width (EW) measurement and variations.

However, before taking any measurements on a series of profiles, it is necessary to prepare them in order to make them comparable: by removing telluric lines, correcting for heliocentric or barycentric velocity, "flattening" the continuum and finally normalizing the spectra.

First step

First of all, telluric lines due to steam molecules within the earth's atmosphere must be removed (H_2O), very present on the $H\alpha$ profile. Several tools are available, depending on the software used. Here, we describe the method using ISIS software.

In this example, we willl use the spectrum of the star beta Cmi. This one shows both the interesting signal, mainly the $H\alpha$ line, but also a multitude of fine absorption lines that "pollute" the profile. It is these atmospheric lines that need to be removed.

Proceedings

Display your spectrum in the tab "**5-profile**".

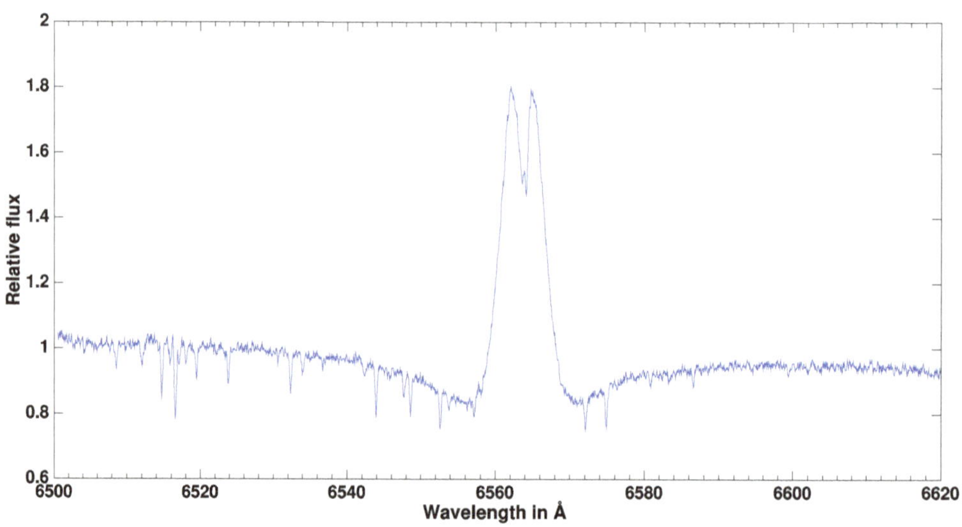

Fig. 5.6.3 — Spectrum of bet CMi. To be corrected for the atmospheric absorption lines seen to the left and right of the $H\alpha$ profile.

Click on the "**H2O**" to the right of the spectral profile.

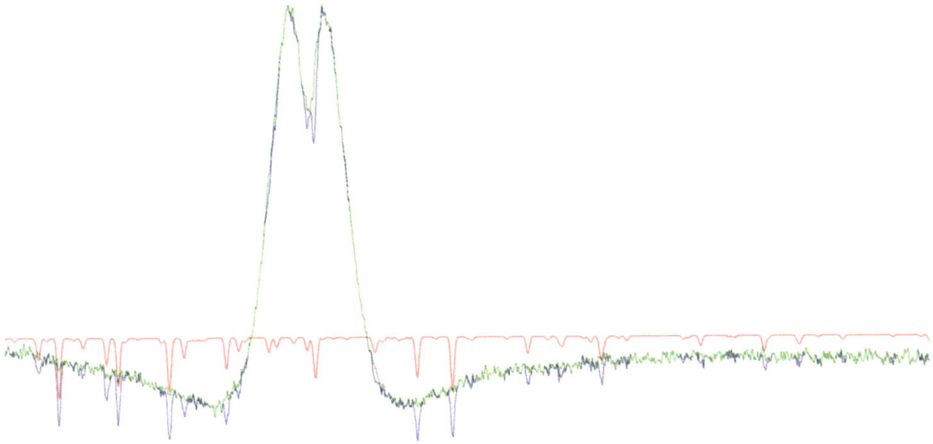

Fig. 5.6.4 — Subtraction of atmospheric lines. Using ISIS software.

The image in Figure 5.6.4 shows the superposition of 3 different profiles:

- In **blue**, the raw spectrum (uncorrected for telluric lines).

- In **red**, the spectrum of water vapor lines calculated according to their theoretical position ("unwanted" signal to be suppressed).

- In **green**, the star's signal "cleaned" of telluric lines. This is the result of dividing your raw spectrum by that of the telluric lines.

We can see that there is a particularly troublesome water vapour line at 6 564 Å located right in the heart of the $H\alpha$ line. To achieve the correct final result (in green), you will need to adjust the 3 parameters in the displayed window (see Figure 5.6.5).

The first cursor in this figure modifies the FWHM of telluric lines. By playing with this slider, you can fine-tune the width of the atmospheric lines with those of your spectrum. This width depends on the resolution of your spectrograph.

The second slider adjusts the intensity of telluric lines which is dependent on the degree of ambient humidity at the time of acquisition. Adjust the amplitude so that you no longer see an absorption line profile on the green curve.

The third slider lets you fine-tune the centring of telluric lines according to calibration precision. If the calibration is not very precise, you will notice it at this stage, as you won't be able to remove all the water vapor lines perfectly. One possible remedy is to repeat the treatment and try to improve it.

Fig. 5.6.5 — Removal settings of atmospheric lines. Using ISIS software.

Please note! Some spectra produced under low humidity conditions (e.g. desert/high altitude) show few or no water vapor lines.

For maximum precision in these parameter settings, which remain highly visual and subjective, you can zoom in on a portion of the spectrum's useful area by clicking on the box at the top right of the spectral profile, and display "**X2**" instead of "**Full**".

When the line adjustment and results appear satisfactory, click on the box "**Close**" written in red. The software will perform the division and display the spectrum without the water vapour lines (see Figure 5.6.6). Depending on the quality of the initial calibration, residual atmospheric lines may remain at the edge of the spectrum, the main thing is not to have any in the profile of the line to be studied. Save this step with a name like "betCMI-woATM" (for this example).

Second step: correct heliocentric velocity

Now it is time to correct each spectrum in your series to the heliocentric or barycentric velocity. The procedure is identical to that described in chapter 3.7 of this book.

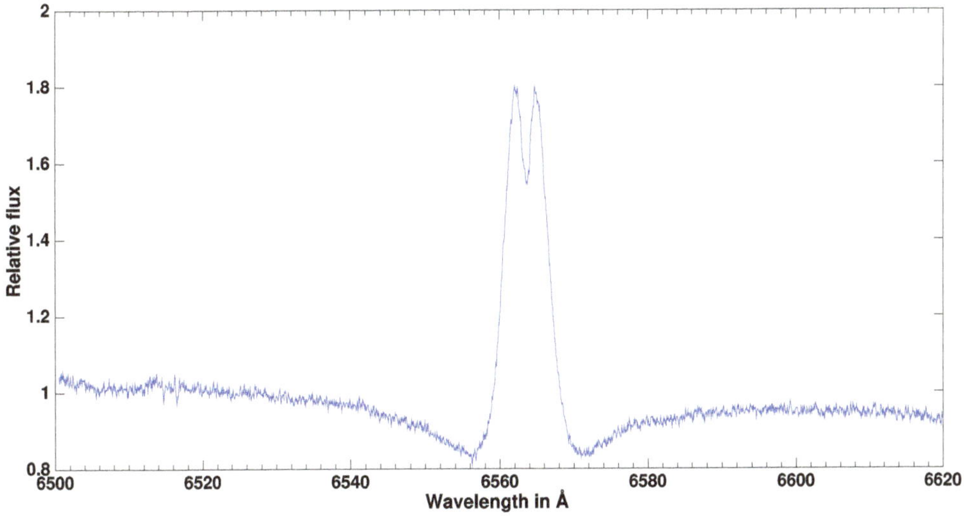

Fig. 5.6.6 — Atmospheric line removal result. Using ISIS software.

Third step: obtaining a "flat" continuum

The aim is to achieve a flat continuum, horizontal over the wavelength range of the line to be measured and of course, to reproduce it identically on all the profiles to be compared. This involves dividing the profile by its own continuum. This is the method to use with ISIS software. Always in the tab "**5-Profile**", click on "**Continuum**" to the right of the spectral profile.

To determine your spectrum continuum, you need to tell the software which zone to exclude, so as not to take the main line into account. In our case, it is the $H\alpha$ line that needs to be deleted (see Figure 5.6.7). Double-click to the left and then to the right of the profile to temporarily delete it, the line is replaced by a line connecting the 2 portions of the star's continuum. Then click on the "**Automatic 1**" case of the window "**Continuum adjustment**". ISIS then displays a blue curve which is the continuum of your spectrum. To complete this step, click on the "**Divide**" case, then close the window. Your spectrum will be displayed with a horizontal continuum.

Fourth stage: normalization

There is one last step to ensure that your spectrum or series of spectra is well prepared for the calculations. They must be normalized on the continuum, set to the value of 1 in the 2 zones to the right and left of the spectral profile.

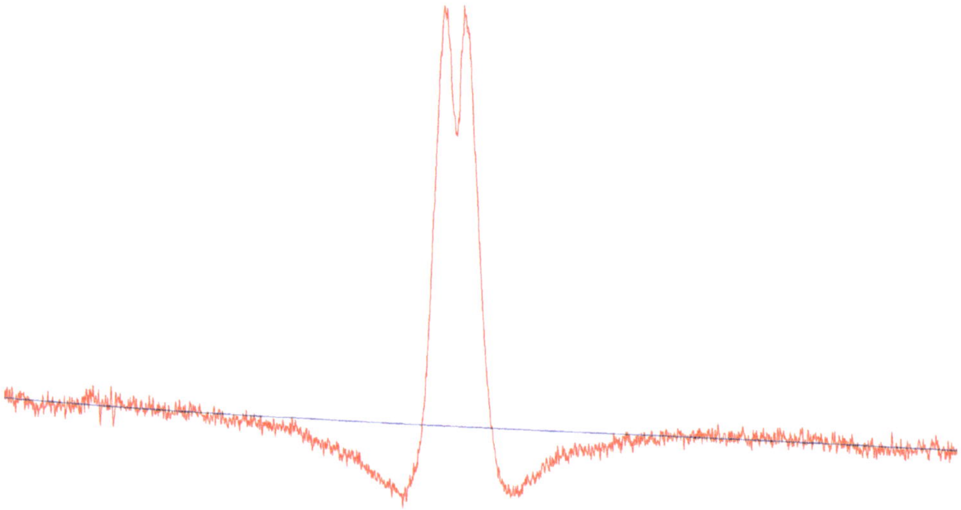

Fig. 5.6.7 — Continuum correction. Using ISIS software.

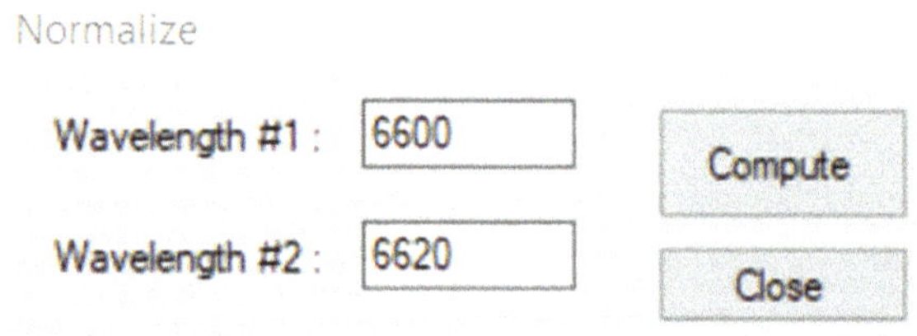

Fig. 5.6.8 — Spectrum normalization window. Using ISIS software.

Click on the box "**Normalize**" to the right of the spectral profile and enter the 2 wavelength bounds for calculating this normalization (see Figure 5.6.8). They must, of course, be identical for the entire series of spectra. In this example, the values 6 600 Å and 6 620 Å to the right of the $H\alpha$ profile were chosen. Close the normalization window and check that the continuum has the value 1 to the right and left of the spectral line.

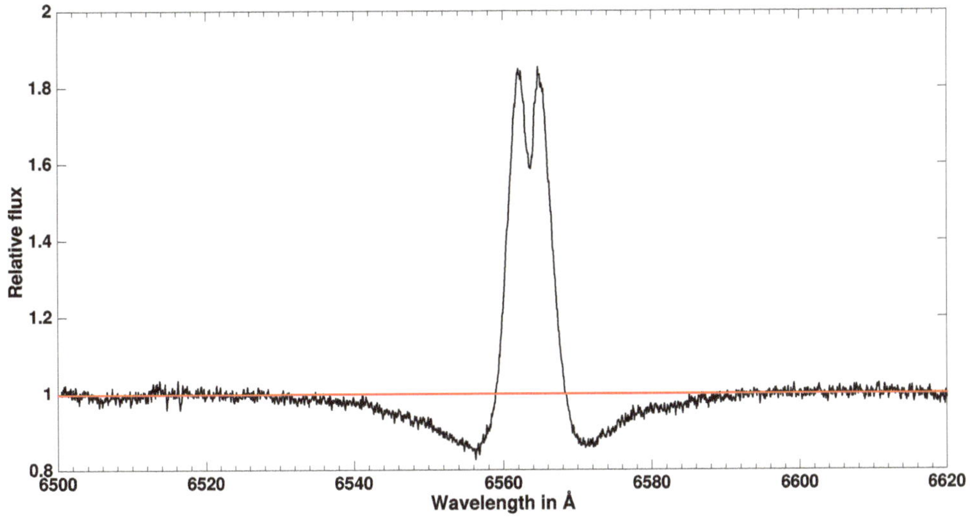

Fig. 5.6.9 — Corrected bet CMI spectrum. Removal of telluric lines, correction of heliocentric velocity, horizontalization of the continuum and normalization to 1.

Of course, the same treatment must be carried out for each spectrum used in the study, in order to obtain a homogeneous, identically calibrated data set. Preparing your data set can take a long time, and it is not uncommon to use a hundred or even a thousand spectra. The BeSS spectral database lets you use stored spectra to carry out your own studies. You can also, if you are familiar with Python programming, create your own processing pipeline for a series of spectra. Several libraries are dedicated to astronomical spectroscopy (astropy, specutils).

To summarize the various corrections we have made in this example (bet CMi):

- Removal of telluric lines from the water vapour present in the region of the $H\alpha$ line.

- Correction of the heliocentric or barycentric velocity of the spectrum.

- Correction to obtain a horizontal continuum by dividing the spectrum by its own continuum.

- Continuum normalized to 1.

5.7 Measuring the V and R peaks of a spectral line

As we have seen, Be stars have spectral profiles that evolve over time. On star profiles showing a double peak, we can sometimes see a shift in intensity from one peak to the other. This variation in intensity or wavelength between the two peaks shows that the disk revolving around the star has a precessional motion that modifies its inclination with respect to the observer's line of sight. Denser zones may also appear during rotation. It is interesting to show these evolutions on a series of spectra and spot whether a periodic phenomenon is manifesting itself.

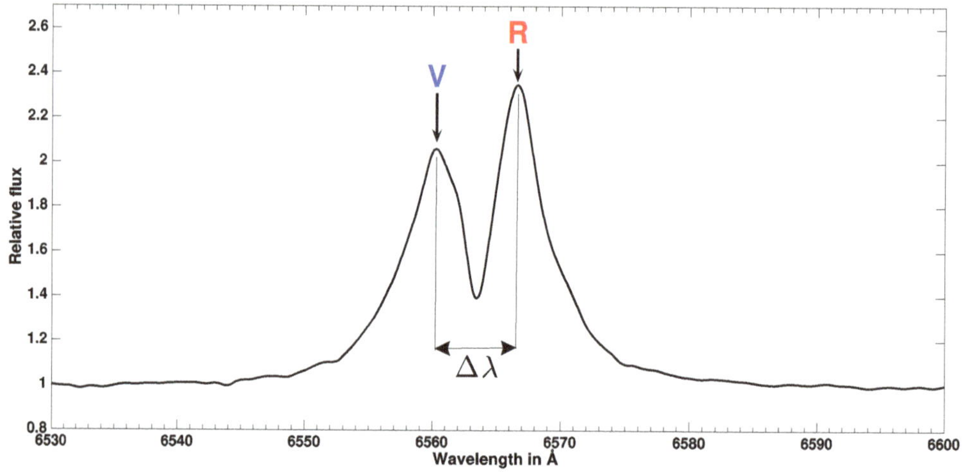

Fig. 5.7.1 — Measurement of V and R peaks. Intensity of the two peaks and wavelength spacing may vary over time. The continuum has been made horizontal and normalized to 1 on this spectrum.

The intensity of the two peaks V and R on each profile in the series is measured using the method described in section 3.8 by fitting a Gaussian curve to each V and R line. We can also measure the $\Delta\lambda$ value between the 2 peaks V and R ($\lambda_R - \lambda_V$), which can show a Doppler-Fizeau shift variation with a velocity ($V \sin(i)$) that varies over time. Other measurements with V and R peaks are possible, In some publications, other calculations using intensity can be found: for example, V+R or V−R.

To have a meaningful representation of developments, a lot of data are required, usually collected over a number of years. The BeSS database can be used to retrieve a large number of spectra on selected targets with selectable search criteria. An uploaded spectrum can therefore be used much later... and the amateurs are sometimes surprised to find themselves quoted in a professional publication. Using a spreadsheet program, we can represent these evolutions by creating a graph with the time in Julian days on the x-axis and the V/R ratio (or V+R or V−R) on the y-axis for each spectrum.

In the same way, we can represent the velocity variation between the two peaks $\Delta\lambda$. These reports are often used in publications.

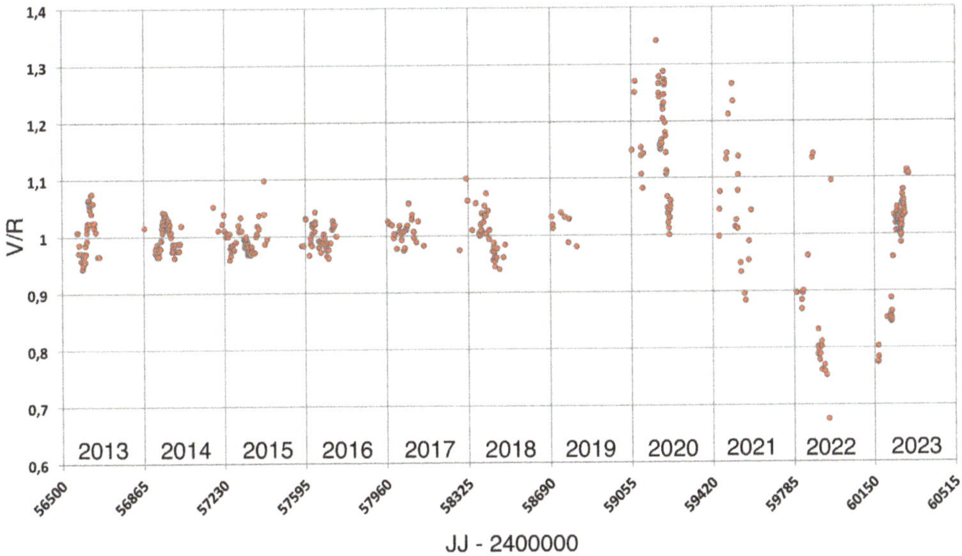

Fig. 5.7.2 — V/R ratio graph for pi Aqr. Based on 400 spectra extracted from the BeSS database for the $H\alpha$ line and covering a 10-year period. On the x-axis, the spectrum date is in Julian days, with the V/R ratio on the y-axis.

We can see that between 2013 and 2019, the V/R ratio is fairly stable, with little difference in amplitude between the 2 peaks, and a V/R ratio close to 1.

It is from 2020 onwards that the amplitude of variations becomes quite significant, with reversals of maximum intensity between the two peaks in subsequent years.

You can also, from the same data, calculate one or more periods. Chapter 10 on spectroscopic binaries describes this procedure.

5.8 Measurement of equivalent width (EW)

The equivalent width (EW) of a line, whether in emission or in absorption, is the measurement of the area that integrates all its flux, in other words the area of the profile between two wavelengths relative to the continuum value.

EW corresponds to the power of the energy emitted or absorbed by a line. (e.g.: abundance of a chemical element) taking into account not only its intensity but also the width of its spectral profile. This surface is reduced to an equivalent rectangle of the same area with a height of 1. Its width is noted as *EW* (for *Equivalent Width*).

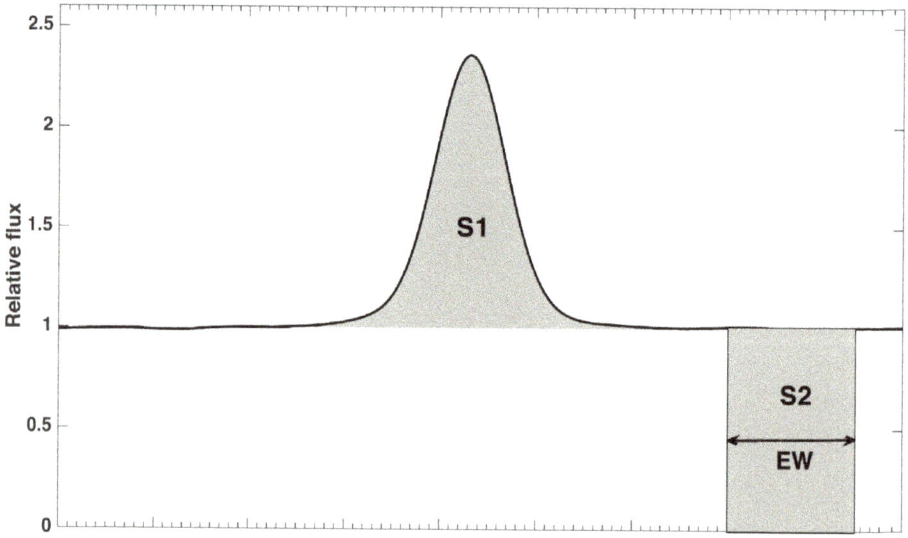

Fig. 5.8.1 — equivalent width *EW* of a spectral line. Areas S1 and S2 are equal, and the equivalent width corresponds to the width of the rectangle expressed in ångströms (Å).

Do not confuse the equivalent width (*EW*) with the full width at half-maximum (FWHM), the first measures a surface area, la second a line width. The mathematical equation for this equivalent width is:

$$EW = \int_{\lambda_1}^{\lambda_2} \frac{I_c - I_\lambda}{I_c} \mathrm{d}\lambda$$

where:

EW: the equivalent width in ångströms (Å).
λ_1 and λ_2: values of the equivalent wavelength width calculation zone that frames the line to be measured.
I_c: the intensity of the continuum.
I_λ: profile intensity at a wavelength λ.

If the continuum is normalized to 1, the equation becomes:

$$EW = \int_{\lambda_1}^{\lambda_2} (1 - I_\lambda)\,\mathrm{d}\lambda$$

Note that if the profile is in emission, the value of the equivalent width (EW) is negative, positive if the profile is in absorption.

Many software programs, such as ISIS and Visual Spec, easily calculate the equivalent width. We will now show a Python script that performs this calculation by delimiting the line to be measured. The spectrum must first be prepared as described in section 5.6. This program is based on a script written by Matthieu Le Lain to display the graph of a spectrum and we have completed with the equivalent width calculation function.

```
import numpy as np
import matplotlib.pyplot as plt
from astropy import units as u
from astropy.visualization import quantity_support
quantity_support()
from astropy.io import fits
from specutils import Spectrum1D, SpectralRegion
from specutils.analysis import equivalent_width
import tkinter as tk
from tkinter import filedialog
root = tk.Tk()
root.withdraw()
spec_path = filedialog.askopenfilename()
file = fits.open(spec_path)
specdata = file[0].data
header = file[0].header
WB=6535 #limite du calcul coté bleu du spectre
WR=6600 #limite du calcul coté rouge du spectre
spec = Spectrum1D.read(spec_path)
EW=equivalent_width(spec, regions=SpectralRegion(WB * u.AA, WR * u.AA))
Min=np.min(spec.flux)
Max=np.max(spec.flux)
FL=np.min(spec.wavelength)
plt.figure(figsize=(15,7), dpi=300)
plt.title(header['OBJNAME'] + ' - ' + header['EXPTIME2']
         + ' - JJ=' + str(header['JD-MID']), fontsize=14)
plt.xlabel('Wavelength in [Å]')
plt.ylabel('Relative flux')
plt.plot(spec.wavelength, spec.flux, 'k')
plt.plot( [WB,WB], [Min,Max], 'b', lw=2 )
plt.text (WB-3,Max-(Max/20), str(WB), rotation=90, fontsize=12)
plt.plot( [WR,WR], [Min,Max], 'r', lw=2 )
plt.text (WR-3,Max-(Max/20), str(WR), rotation=90, fontsize=12)
plt.text (FL,((Min+Max)/2), "EW = {0:.2f}".format(EW), fontsize=16)
nom=header['OBJNAME'] + '_' + str(header['JD-MID'])+'_EW-graph.pdf'
plt.savefig(nom)
plt.show()
```

Fig. 5.8.2 — Python script for calculating the equivalent width.

Before executing the script shown in Figure 5.8.2, the calculation limits for the equivalent width must be entered: λ_1 and λ_2 in wavelength in WB variables (wavelength at left of profile: blue) and WR variables (wavelength at right of profile: red) in ångström (Å) on lines 17 and 18 of the script.

When you run the script, a window appears to select the spectrum you wish to calculate. The example in Figure 5.8.3 was performed on the $H\alpha$ line of a spectrum of the star γ Cas.

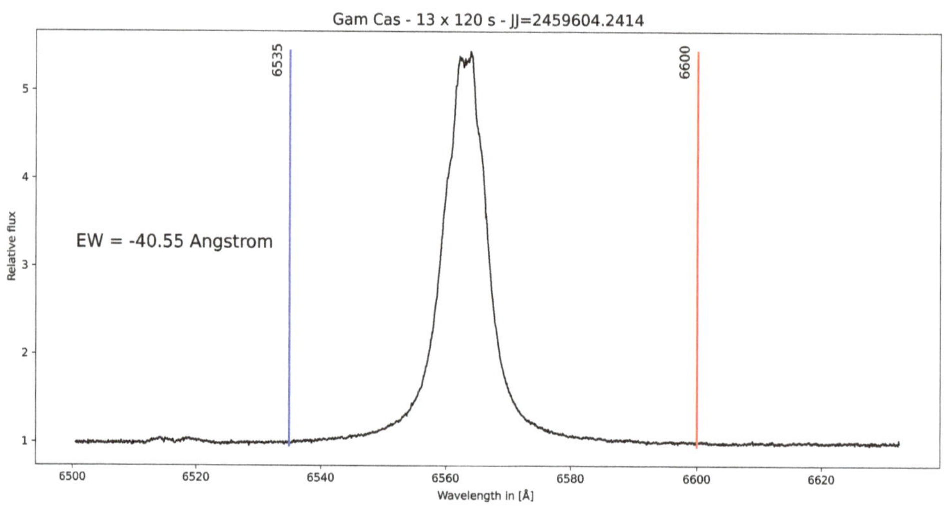

Fig. 5.8.3 — Equivalent width *EW* of the H alpha line of gam Cas star. Boundaries must be selected with precision, integrating the entire spectral profile until it merges with the continuum. The spectrum continuum has been made horizontal and normalized to 1.

The result is displayed with the spectral profile as well as the calculation limits "red" side and "blue" side of the spectrum, which shows the portion taken into account when calculating this equivalent width.

If you have a large series of spectra to analyse, spectrum-by-spectrum calculations can be tedious and time-consuming, you can adapt the Python script according to the data you wish to process. Then, by grouping the results in a spreadsheet, you can plot the equivalent width as a function of date.

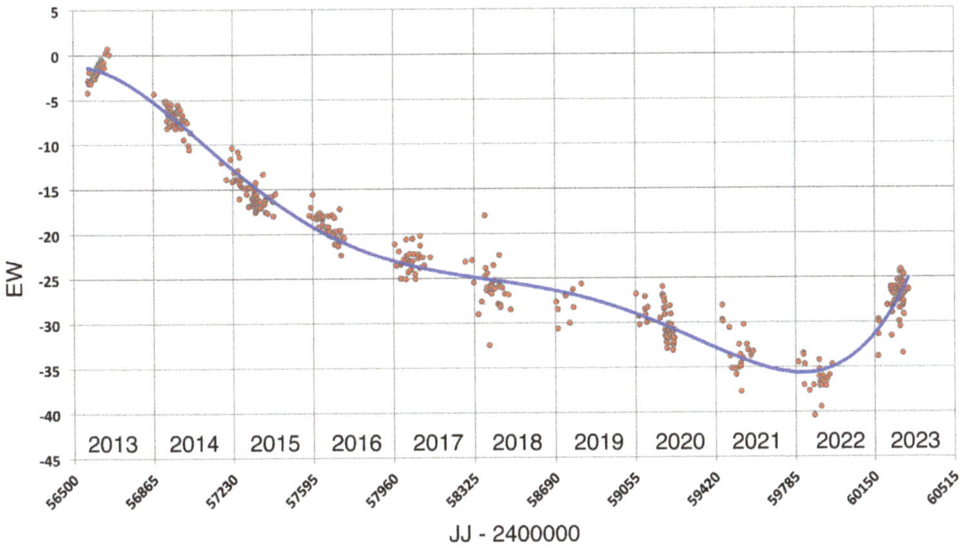

Fig. 5.8.4 — Equivalent width graph *EW* of Pi Aqr. Using 400 spectra extracted from the BeSS database: $H\alpha$ line over 10 years. On the x-axis, the spectrum date in Julian days. On the y-axis, the equivalent width EW in ångströms (Å).

This graph shows changes in equivalent width over 10 years. This series of spectra was processed in the same way. The blue curve is the measurement trend of the scatterplot. In the case of pi Aqr, we seem to have reached a maximum in 2022, with the curve for 2023 tending to decrease in intensity.

Measurements over long periods are time-consuming but provide important information about the long-term evolution of a star. Each observer can produce their own long-term data. It can also be completed using spectra from the BeSS database. The advantage of producing your own data is that your spectrograph is always set up in the same way and you can always apply the same processing method with the same software, to obtain homogeneous, reproducible data. The observer may well plan to regularly monitor over several years, 4 to 5 Be stars, measure the spectral evolution of this sample and make the results available to the scientific community through a personal website or astronomical forums and discussion lists.

5.9 Bibliography

- Pulsation, rotation, wind and magnetic field in early B-Type stars by Coralie Neiner, 2002. ISBN 90-5776-090-8.

- Astronomie-Astrophysique, Introduction by Agnès Acker, Dunod, 2005. ISBN 2-10-005462-7.

- Astronomie: méthodes et calculs by Agnès Acker and Carlos Jaschek, Masson, 1995. ISBN 2-225-84689-8.

- Astronomical Spectrography for Amateurs: collective of authors, EDP Sciences, 2011. ISBN 978-2-7598-0629-4.

- Numerical Python in Astronomy and Astrophysics by Wolfram Schmidt and Marcel Völschow, Springer, 2021. ISBN 978-3-030-70346-2.

- Spectroscopic properties of stars in young binaries fundamental data for understanding binary formation and disk evolution by Lisa Prato, the European Physical Journal Plus, Vol. 138, article number 244, 2023.

- Equivalent width measurements, spectral properties and evolution of local cluster galaxies by J. Fritz et al., Astronomy and Astrophysics A 566, A32, 2014.

- On the Variability of the Central Absorption in the $H\alpha$ Line Profile of ζ Tau by Ernst Pollmann, SSRN, 2022.

- V/R variability and global oscillation in Be star disks by R.W. Hanuschik, W. Hummel, O. Dietle and E. Sutorius, Astronomy and Astrophysics 300, 163-176, 1995.

- V/R variations in $H\beta$ emission profiles of Be stars by R.E. Mennickent and N. Vogt, Astronomy and Astrophysics, 241, 159–166, 1991.

- The Role of Disk Tearing and Precession in the Observed Variability of Pleione by K. C. Marr et al. The Astrophysical Journal, 928, 145 (15 pp), 2022 April 1.

- Cyclic variability of the circumstellar disk of the Be star ζ Tauri by S. Stefl, Th. Rivinius *et al.* Astronomy and Astrophysics, 504, 929–944, 2009.

- The use of terms and definitions in the study of Be stars par George W. Collins II. Cambridge University Press, 1987. DOI: 10.1017/S0252921100115969.

- Doppler tomography of the circumstellar disk of Pi Aquarii by S.V. Zharikov *et al.* (Astronomy and Astrophysics 560, A30, 2013.

- Evolution of the disk of Pi Aqr from near-disappearance to a strong maximum by Yaël Nazé et al., New Astronomy, Vol. 73, November 2019, 101279.

- Outbursts and stellar properties of the classical Be star HD 6226 by N. Richardson *et al.*, MNRAS 508, 2002–2018, 2021.

- Star and their spectra, an introduction to the spectral sequence by James B. Kaler, Cambridge University Press, 2011. ISBN 978-0-521-89954-3.

- Stellar Spectral Classification by Richard O. Gray and Christopher J. Corbally, Princeton series in Astrophysics, 2009. ISBN 978-0-691-12510-7.

- The astrophysics of emission-line stars by Tomokazu Kogure and Kam-Ching Leung, Springer, 2010. ISBN 978-1-4419-2236-6.

- BeSS Monthly reports from BeSS contributors written by Valérie Desnoux: http://valerie.desnoux.free.fr/BeSS/index.html

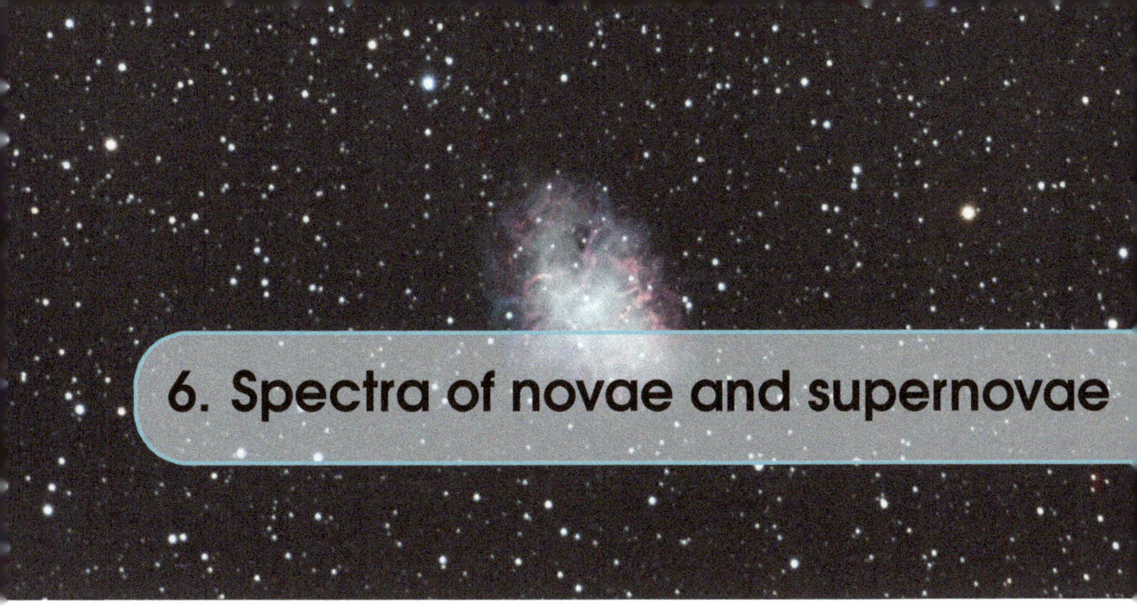

6. Spectra of novae and supernovae

Novae and supernovae are very interesting phenomena for spectroscopic observation. These two types of objects sometimes become visible in the sky to the naked eye for a period of a few weeks to a few months due to the sudden increase in their luminosity following a gigantic explosion. This is one of the areas of observation where we need to be very reactive and be able to produce a spectrum as quickly as possible after the announcement, typically within a few hours. Obtaining a spectrum as close as possible to its discovery enables us to classify the type of object and will provide a wealth of information on the nature and physics of the phenomenon. Regular monitoring during the first few days is also important to observe its evolution.

6.1 Nova or supernova?

Although their respective names might suggest that they are of the same nature, these two types of object are very different. The main difference is that in a nova, the star (a white dwarf) survives the explosion, whereas in a supernova, the whole star explodes (the original star and the resulting remnant is different depending on the type of supernova). Historically, to signal a "new star" in the sky, the Ancients spoke of "*Nova stella*" (New Star in Latin). They thought she had just been born... The Chinese spoke of "the guest star". For several months, even in broad daylight, they observed the 1054 supernova that gave rise to the Crab pulsar and the beautiful M1 nebula. The word "nova" stuck, and "supernova" for the most powerful, but that was before we understood the phenomena involved.

Novae and supernovae are not new stars, they just become brighter when they explode, making them easier to spot. Each has a very different origin and physics, which we will describe in this chapter. The energies involved are very different. The absolute magnitude of a supernova is much greater than that of a nova. Absolute magnitude for type 1a supernovae is around -19.25 are all relatively consistent. They can therefore be used as standard candles.

6.2 Where to find information?

To be able to be reactive and produce a spectrum quickly, as soon as possible after the discovery of a nova or supernova, you need the information. It can be found on many websites. Some of these sources are presented here, most widely used at the time of writing this book. What is more, they are free and open to all.

The website "Transient Name Server":

https://www.wis-tns.org/

This is a "warning or whistle-blower" site reporting transient astronomical phenomena such as supernovae candidates (PSN = possible supernova), but also everything that unexpectedly becomes brighter (outbursts, novae, variable stars, etc.). It is also a database of spectra/photometry of reported objects: SN, classical novae, etc. You can register on the website to receive e-mail alerts and you can also upload your spectra.

Please note: Normally, this site is for classification purposes only, one spectrum per object is sufficient. But it is not forbidden to file more than one. Note though that this facility is mainly for use by professionals so only use it if you are confident in the quality of your spectra and your classification. This allows us to see the often spectacular evolution over time.

Note: The function "**search**" accessible without registration on the home page allows you to specify queries and create a list that can be sorted by date, magnitude, height, etc.

The website "The Astronomer's Telegram" also provides information on various objects and alerts in near real time:

https://astronomerstelegram.org/

All telegrams can be consulted online, but you can also subscribe to receive emails according to your criteria by selecting the options (for example nova or supernova).

The AAVSO (American Association of Variable Star Observers):

https://www.aavso.org/

This more than 100-year-old association, has a website, database and forum, as well as numerous online tools for generating a field map on demand, trace the light curve of a variable star, and much more.

By registering on this site, you can receive information on astronomical news.

More generally, all spectroscopic discussion lists and astronomy forums where there are sections dedicated to spectroscopy and photometry, can provide you with useful information.

6.3 Novae

A nova is a thermonuclear explosion in an envelope of accreted matter on the surface of a white dwarf in a binary system. The donor companion may be a main sequence star of late spectral type, M most of the time, in the case of classic novae, or a red giant in the case of symbiotic novae. The strong gravity of a white dwarf attracts the matter emitted by its companion, most often via an accretion disk. The envelope becomes denser and denser, and the pressure and temperature rise until the conditions necessary for a thermonuclear reaction are reached.

As the explosion begins, the nova's brightness will increase sharply by as much as 10 magnitudes. A large part of the gaseous envelope is expelled at velocity of several hundred to thousands of kilometers per second. The nuclear reaction of the rest of the gaseous envelope continues for several months or years (depending on the characteristics of each nova) by emitting a powerful stellar wind. White dwarf radiation excites the gas in the envelope, and emission lines can be seen in the spectrum.

Then over time, the ejected envelope becomes increasingly diluted and nebular lines (forbidden lines) begin to appear in the spectrum. Brightness decreases and the nova becomes difficult to observe at magnitudes greater than 17.

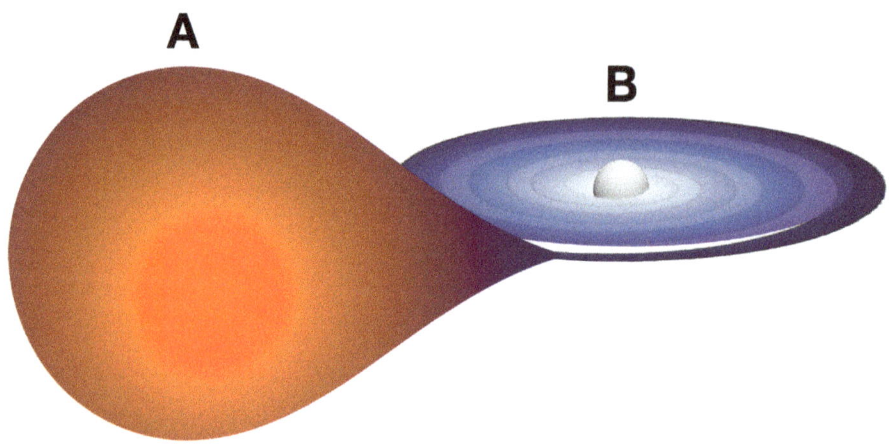

Fig. 6.3.1 — The structure of a nova. Component A is a cold, sparsely populated star, while component B is a white dwarf. The material of the donor star is attracted by the gravitational force of the white dwarf, forming a gas/dust accretion disk around it. Some of the disk material is transferred to the envelope surrounding the white dwarf.

Novae have been observed for over a century, whether in spectroscopy or photometry. Some of them show explosions at regular intervals such as U Sco, T Crb, V745 Sco, T Pyx and RS Oph stars. We call them recurrent novae. Hence the importance of regular follow-up between two outbursts.

Novae behave differently depending on various parameters such as the mass of the white dwarf. The phenomenon can evolve more or less rapidly. Various systems have been proposed in an attempt to classify them according to the speed of extinction.

The first system dates back to 1936 by B. P. Gerasimovic offering 2 classes: "flashing" and "slow".
Other classifications were later proposed: Dean B. Mc Laughlin in 1939 and 1945, then C. Bertaud in 1951.

The most detailed and probably the most widely used to date is that produced in 1957 by the Anglo-American astronomer Cecilia Payne-Gaposchkin from Harvard University.

She proposes a classification of novae into 5 classes according to their speed of evolution. She indicates for each class the time required to observe a decline of 2 and 3 magnitudes since the maximum brightness measured.

Class	t_2 (days)	t_3 (days)
Very fast	< 10	< 20
Fast	11-25	21-49
Moderately fast	26-80	50-140
Slow	81-150	141-264
Very slow	151-250	265-440

Fig. 6.3.2 — Table of classic novae velocity classes. (According to Cecilia Payne-Gaposchkin in 1957). The columns t_2 and t_3 indicate the time in days to see an increase in magnitude of 2 and 3 respectively since the maximum brightness.

We can see that, depending on the novae, the speed of decay is highly variable, ranging from a few weeks to several months or years.

According to R. Williams (1990 and 1992), we can distinguish 3 classes of nova spectral line evolution other than the hydrogen lines, which remain the most intense in emission:

• Fe II novae: these show intense lines of ionized iron once (Fe II) in emission in the ejecta and indicate expansion velocities of between 1 000 and 3 500 $km.s^{-1}$. They make up about 60 % of all novae.

• He/N novae: these show intense lines of neutral helium (He I), of ionized helium once (He II) and ionized nitrogen once and two times (N II, N III). Ejecta velocities are in excess of 2 500 $km.s^{-1}$. They account for around 30% of all novae.

• The other novae, about 10% of the total, which don't include any forbidden lines, are in a class of their own.

This classification has often been used in past scientific publications but it is now obsolete (Aydi *et al.*, 2023): the appearance of the Fe II and He/N lines corresponds to the evolution of the nova's spectrum, not its type (see for example Steve Shore, 2013).

The nova phenomenon is not exceptional, it happens frequently in our galaxy, but novae can also be seen in nearby galaxies such as M31, or the two Magellanic clouds.

Overall, typically 3 to 4 novae are bright enough to be observed spectroscopically every year, but it is quite rare to see very bright ones (magnitude < 7-8), accessible at medium and high spectral resolution.

A study of the probability of discovering classical novae according to their magnitude has been created by Hilmar W. Duerbeck in 1990 based on observations made over a century.

Mag. V maxi	Probable appearance (Years)	Mag. V maxi	Probable appearance (Years)
< 1	25	4-5	20
1-2	50	5-6	7.15
2-3	25	6-7	2.13
3-4	33.33	7-8	1.72

Fig. 6.3.3 — Table of novae discoveries over a century. According to Hilmar W. Duerbeck in a book by Michael F. Bode and Aneurin Evans, *Classical Novae*, second edition. Cambridge University Press (2008). ISBN 978-1-107-40534-9.

One of the brightest novae of recent years was nova V339 Del, discovered in 2013. Its magnitude $V = 6.5$ allowed a large number of amateurs to follow it with the production of over 1 000 spectra over a period of a few months. We'll talk more about this later.

The spatial distribution of novae lies mainly between +15 and -15 degrees around the galactic equator and closer to the center of the Milky Way, due to the greater concentration of stars in these areas. We also note that novae are found preferentially in areas of low interstellar extinction (Plaut 1965). Shara and Drissen reported in a 1995 publication that a nova was observed in the globular cluster M80 in 1860.

6.4 Observing the spectrum of a nova

When you have decided to make a nova spectrum, the first question to ask is what resolution and what spectrograph to use?

The advantages of using a spectrograph like the UVEX, Star'Ex or LHIRES III, is to be able to easily change the grating and therefore the resolution. But unfortunately, if the target brightness is low, it is not always possible to obtain a good high-resolution spectrum, it will certainly be very noisy, making it difficult to identify fine line changes.

In this case, the low resolution will still provide very important information, including in the near UV or near IR range with a UVEX. You therefore need to know the limiting magnitude of your observation configuration, depending on the resolution you have chosen.

6.4.1 Low-resolution spectra

As soon as the nova's magnitude is > 9-10, it becomes difficult to produce a high- or medium-resolution spectrum.

This requires the use of low resolution spectrographs like the Alpy 600 or the LISA. You can also use instruments with interchangeable gratings, such as the LHIRES III, the UVEX or the Star'Ex, using a grating of 300 to 600 lines/mm.

The advantage of the 300 lines/mm grating is that it covers the entire visible range from around 400 to 750 nm, or even a little more, depending on the spectrograph and camera used.

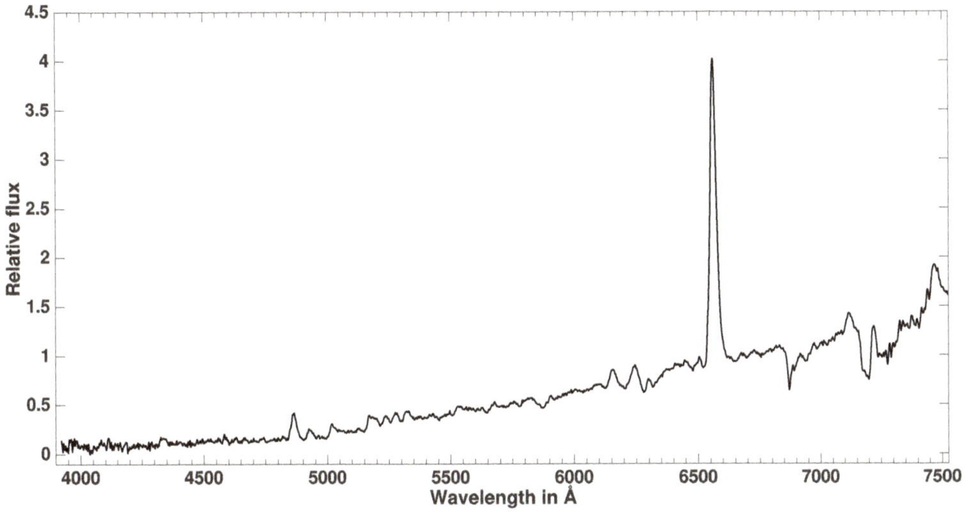

Fig. 6.4.1 — V5853 Sgr spectrum (Nova Sgr 2016b). Spectrum of the entire visible range, produced with a LISA spectrograph (equipped with a 35 µm slit) with an ATIK 414EX camera on a C14. Resolution $R = 780$, 10 exposures of 600 seconds. Note the strong emission on the $H\alpha$ line. On the x-axis, wavelength in ångströms. On the y-axis, intensity relative to the continuum normalized to 1.

Using a UVEX, you can also explore the near-IR and near-UV sides of the spectrum, which will bring additional information to the "visible" spectrum. As always, we recommend observing a nearby reference star at the same height as the nova, in order to obtain a good instrumental response.

6.4.2 Medium/high resolution spectra

If the nova's magnitude gives access to medium or high resolution, you can't afford to miss out on this opportunity. The tricky part, however, is deciding which wavelength to observe, as many lines evolve very quickly. A good place to start is with the $H\alpha$ line, which is easy to observe and spectacularly intense. Better still! It only grows over time.

Spectra can then be produced at other wavelengths as the nova evolves and the need to obtain a spectrum for a particular wavelength. With a LHIRES III or Star'Ex HR spectrograph (equipped with a 2 400 lines/mm grating), set up the spectrograph "on table" with the aid of daylight, so as to center the line to be studied. This centering is important because the lines can be very wide or even extend beyond the camera's sensor. If you're lucky enough to have an Echelle-type spectrograph like the eShel or BACHES, there is a definite advantage to using it because these instruments cover the entire visible spectrum with a resolution $R > 9\ 000$ in a single acquisition.

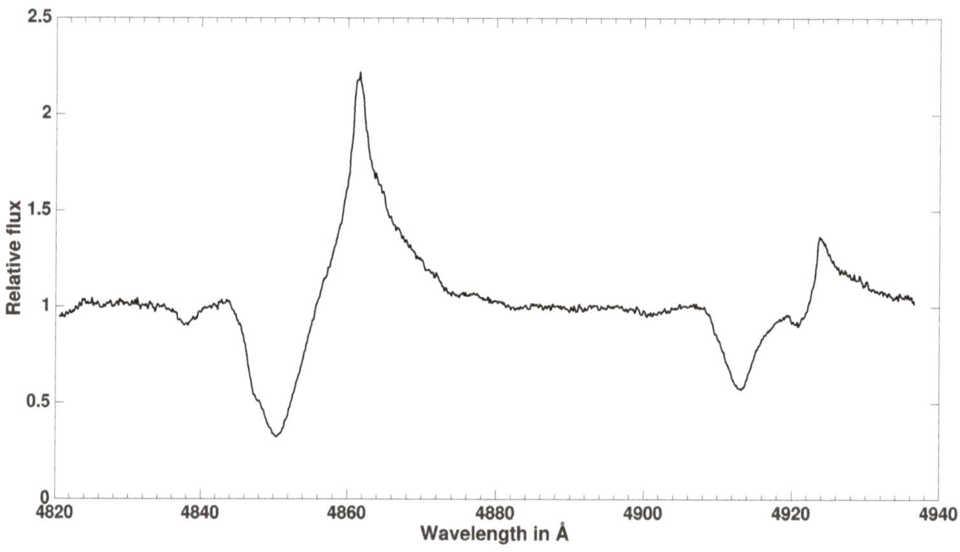

Fig. 6.4.2 — Spectrum of V1405 Cas (Nova Cas 2021). Centred on the $H\beta$ line on the left and He I 4 922 Å on the right. The 2 lines present a p Cyg profile. Realized with an eShel spectrograph on a 400 mm Ritchey-Chrétien telescope and an ATIK 460EX CCD camera. Resolution $R = 11\ 000$, 16 exposures of 300 seconds. On the y-axis, intensity relative to the continuum normalized to 1.

According to the evolution of the nova's magnitude, you can also switch spectrographs from medium/high resolution to low resolution. The important thing is to be able to follow the evolution of the spectrum for as long as possible. This evolution will be more significant in the first few days after the nova appears, when a daily spectrum is required, then you can space out the observations depending on the speed of evolution, to produce just one or two spectra per week, even just one a month when in the nebular phase where developments are very slow. As each nova is a special case, you need to adapt your observations according to photometric trends.

Novae spectra can be submited to the ARAS database:

https://aras-database.github.io/database/index.html

It is a way of centralizing the spectra produced which can then be used by professionals. It was created by François Teyssier and Jaroslav Mercand has been running for many years now. In addition to novae spectra, this database also hosts spectra of symbiotic stars, cataclysmic and dwarf novae. To submit a spectrum in this database, send it in .fit or .fits format to one of the 2 following addresses:

francoismathieu.teyssier@gmail.com

arasdatabase@gmail.com

On receiving your spectrum, François Teyssier will check and validate it in the database, using the information in the .fit file header. From this database, you can also download spectra that interest you and compare them with your own.

6.5 The P Cyg profile

In novae spectra, you will often encounter profiles of the "P Cyg" type. A profile like this is the signature of an expanding mass of matter around a star or the presence of stellar wind. It is named after the star P Cyg (34 Cyg) well known for its powerful wind, where it was first observed.

This profile is characterized by the presence of an emission peak, centered on the line to be studied, preceded by a "hollow" in absorption and blue-shifted showing a mass of ejected matter approaching in the observer's line of sight. This part of the absorption spectrum also gives an indication of the opacity of the ejecta. It is possible to calculate its expansion velocity.

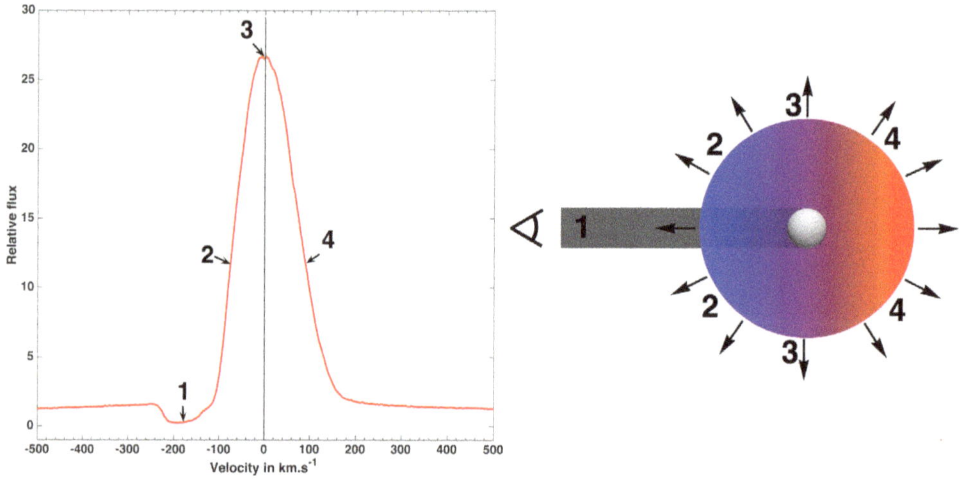

Fig. 6.5.1 — Type P Cyg profile. Peak [3] represents the material ejected by the star along the observer's line of sight, on the left flank [2], we can see the material approaching the observer and away from it on the right flank [4]. The absorption part [1] corresponds to the area where ejecta is in front of the star directly along the observer's line of sight.

We often represent the graph of such a spectrum in velocity, not in wavelength as this provides direct information on the speed of the ejecta or stellar wind (in the case of a spherical ejecta).

6.6 Converting a wavelength graph into a velocity graph

To convert a graph whose abscissa (the x-axis) is in wavelength to a graph in velocity, we calculate the Doppler-Fizeau shift of the spectrum relative to the value of a line at rest, for example $H\alpha$ or HeI 5 876 Å, by using the following equation:

$$Vr = \frac{\lambda - \lambda_{Ref}}{\lambda_{Ref}} c$$

where:

Vr: the velocity in km.s^{-1}.
λ: the wavelength to be converted into velocity.
λ_{Ref}: the reference wavelength, at rest, i.e. zero velocity, for example on the $H\alpha$ line = 6 562.82 Å.
c: the speed of light = 299 792.458 km.s^{-1}.

We will describe two methods for obtaining a spectrum calibrated in velocity with respect to a particular line.

6.6.1 Using ISIS software

There is a tool available in ISIS for converting a wavelength-calibrated 1D .fits file into a velocity-calibrated .dat file. This function is accessed via the "**Misc**" tab, then "**Heliocentric velocity**", finally, in the frame on the right "**Wavelength to Velocity conversion**".

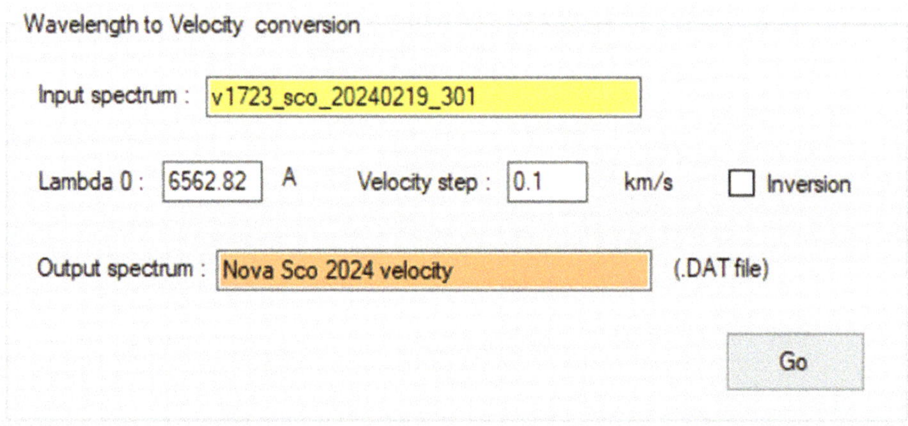

Fig. 6.6.1 — Wavelength to velocity conversion window. From ISIS software.

Where:

• Input spectrum: the name of the .fits or .fit file to be converted to velocity.

• Lamba 0: the value in ångströms of the reference line. In this example, it is the value of the alpha hydrogen line.

• Velocity step: the interval between each velocity measurement point in the destination file.

• Output spectrum: the name of the file converted in velocity and in .dat format.

To display the converted spectrum, you can use a spreadsheet or the ISIS graph printing function (with the Gnuplot interface).

6.6.2 With a Python script

If you are familiar with this language, here's a script that displays the graph of a velocity-calibrated spectrum from a reference line (where $V = 0$).

```
1 import matplotlib.pyplot as plt
2 from astropy import units as u
3 from astropy.io import fits
4 import astropy.wcs as fitswcs
5 from specutils import Spectrum1D, SpectralRegion
6 import tkinter as tk
7 from tkinter import filedialog
8 root = tk.Tk()
9 root.withdraw()
10 line=6562.82
11 Line_name=r'$H\alpha$'
12 spec_path = filedialog.askopenfilename()
13 file = fits.open(spec_path)
14 specdata = file[0].data
15 header = file[0].header
16 spec = Spectrum1D.read(spec_path)
17 lamb=spec.wavelength/u.AA
18 velocity=((lamb-line)/line)*299792.458
19 plt.figure(figsize=(6,4), dpi=300)
20 plt.plot(velocity, spec.flux)
21 plt.title(header['OBJNAME'] + ' ' + Line_name + ' Velocity')
22 plt.xlabel('Velocity in $km.s^{-1}$')
23 plt.ylabel('Relative flux')
24 plt.minorticks_on()
25 plt.grid()
26 nom=header['OBJNAME'] + '_Velocity.pdf'
27 plt.savefig(nom)
28 plt.show()
```

Fig. 6.6.2 — Python script for a velocity graph.

In the variable on line 10 (see Figure 6.6.2), indicate the value in ångströms of the reference line in relation to which the velocity is to be calculated. Line 11 is used to enter the name of this line, which will be displayed on the graph. Then run the script, which will ask you to select the spectrum to be transformed (1D in wavelength-calibrated .fit format). The output is a graph of the same spectrum but calibrated in velocity ($km.s^{-1}$) saved as a .pdf file.

The graph shown in Figure 6.6.3 was generated by the Python script in Figure 6.6.2. The presentation can be improved by adjusting the limits on the abscissa according to the velocity measured or by changing the display of the graph's background grid.

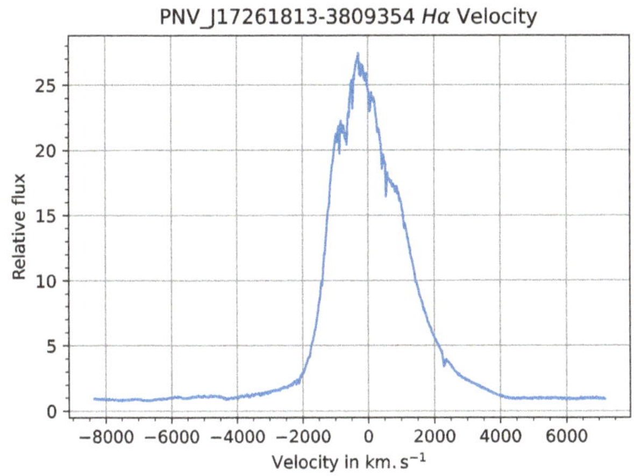

Fig. 6.6.3 — Nova Sco 2024 spectrum centred on the $H\alpha$ line. The original name of this nova was PNV J17261813-3809354. Following confirmation by the International Astronomical Union (IAU), the object is now known as V1723 Sco (variable 1723 of Scorpio). On the *x*-axis, velocity in km.s^{-1}. On the *y*-axis, relative intensity with respect to the continuum normalized to 1.

6.7 A practical case study with the nova V339 Del

On august 14th, 2013, we were lucky enough to see a nova appear in the Dolphin constellation, well placed in the summer sky for all-night viewing. Its magnitude of 6.5 at the time of discovery, made it a prime target for medium and high spectral resolution.

Fig. 6.7.1 — Position of the nova V339 Del. On the left, the nova's position at the time of its discovery, on the right, an image of DSS2 survey taken with Aladin software.

It was discovered by a Japanese, Koichi Itagaki, August 14 at 14 h UT with a 180 mm diameter telescope. He immediately informed the astronomical community, which enabled other observers to react quickly, informed by on line mailing lists and forums.

V339 Del's coordinates are:

RA: 20h 23' 30.68" (J2000)
DEC: +20 46 03.77 (J2000)

The first spectrum was recorded on the day of its discovery at 19 h 45 UT, just under 6 h after detection. Its magnitude of $V = 6.5$ made it possible to produce medium- and high-resolution spectra and many amateurs were able to follow it day after day. It even became brighter during the first 2 days, peaking at $V = 4.45$ on August 16, 2013. It was then visible to the naked eye.

From September 1, 2013 it became fainter than magnitude $V = 7$, but medium- and high-resolution spectra could still be made, and amateurs were able to produce such observations until the end of December 2013.

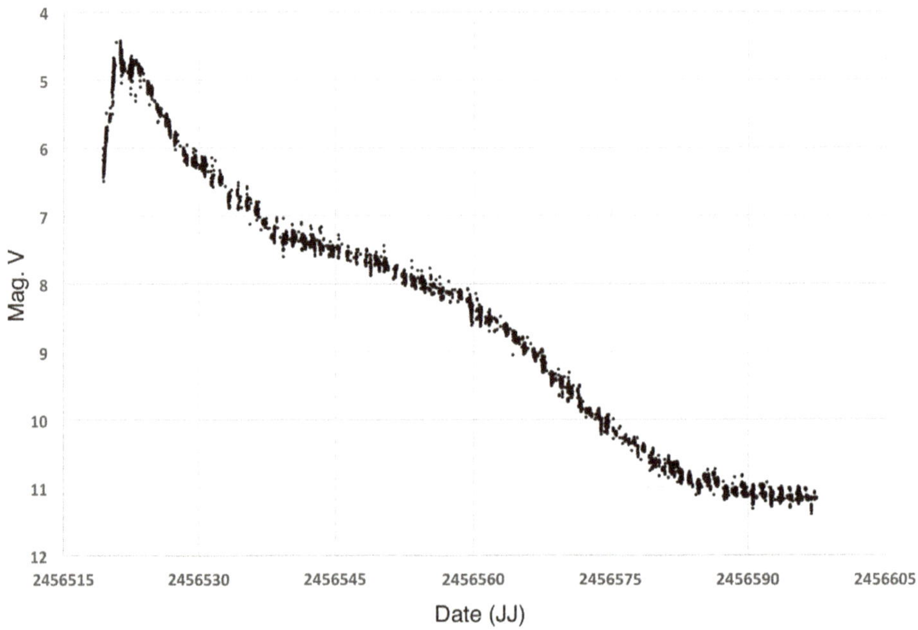

Fig. 6.7.2 — Luminosity curve for nova V339 Del. Realized with magnitude V measurements available in the AAVSO photometric database.

You can consult the large number of amateur spectra in the ARAS database by following this link:

https://aras-database.github.io/database/novadel2013.html

Here is a description of 3 important phases in the evolution of this nova. There are, of course, other stages and physical phenomena involved, but it would take more than a book to describe everything that can be seen in the spectrum over time.

All these evolutionary phases are very well explained in an article by Steven N. Shore published in 2012: "*Spectroscopy of Novae: a user's manual*" which can be downloaded by following this link:

https://arxiv.org/pdf/1211.3176.pdf

The "fireball" phase

This is the very first stage of the nova, during which matter is ejected at speeds of several hundred to several thousand of km.s^{-1}. Hydrogen lines all show a P Cyg profile, as do some helium lines (He I 6 678Å). After a few days, we can already see that the P Cyg profile changes significantly with an increase in the intensity of the emission line and the evolution of the blue part of the absorption profile.

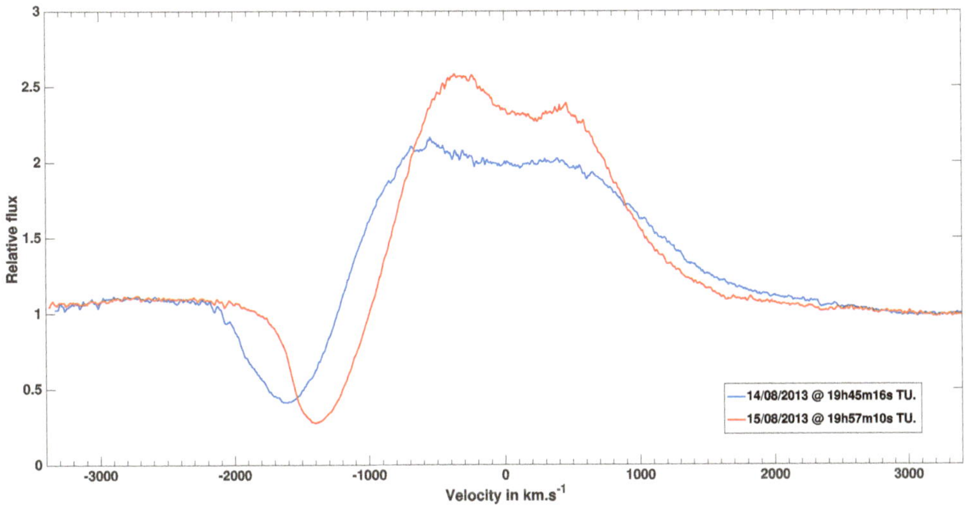

Fig. 6.7.3 — Evolution of the $H\alpha$ line of the nova V339 Del over 24 hours. Spectra made with an eShel spectrograph and an ATIK 460EX CCD camera on a C14 at 24-hour intervals between August 14 and 15, 2013. On the x-axis, wavelength in ångströms. On the y-axis, intensity relative to the continuum normalized to 1.

The graph shown in Figure 6.7.3 shows the evolution of V339 Del in just one day. The expansion velocity of the ejecta for the $H\alpha$ line is of the order of 2 400 km.s^{-1}. We can also see less opacity in the absorption part of the August 15 spectrum. It will disappear within a few weeks as the nova becomes optically more "transparent". By producing spectra day after day, we see an intensification of the emission lines and greater transparency of the ejecta as it dilutes in space.

The "iron curtain" phase

At this stage of the evolution, in the case of V339 Del which is a Fe II-type nova, we are in a cooling phase. We then see the emission lines of ionized iron develop. These Fe II lines are visible from the very first days of the nova's appearance but their intensity will increase more rapidly from the 7th day after the explosion.

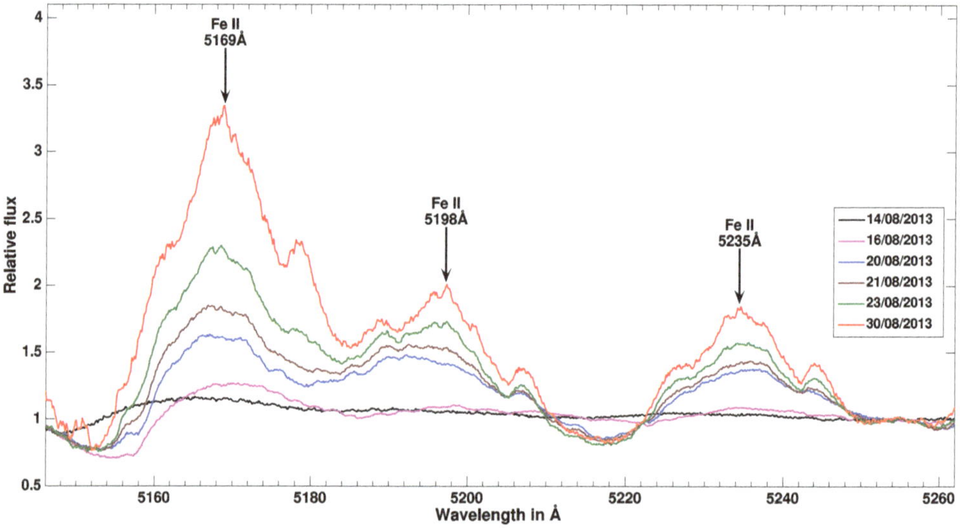

Fig. 6.7.4 — Fe II line evolution. Spectra made with an eShel spectrograph and an ATIK 460EX CCD camera on a C14 between August 14 and 30, 2013. On the x-axis, wavelength in ångströms. On the y-axis, intensity relative to the continuum normalized to 1.

There are other Fe II lines such as those at 4 924 Å or 5 018 Å; we show here those that are most closely grouped together. In order to center these lines on a single spectrum, position the Fe II 5 198 Å line towards the middle of the sensor.

On August 30, the date of the most intense Fe II peak on the graph, the magnitude of V339 Del was $V = 7.1$.

The nebular phase

After a few months, V339 Del reduces futher in brightness and slowly decreases from $V = 11$ in January 2014 to $V = 17$ in September 2023 (photometric data from AAVSO). During this long phase, we find emission lines similar to those of a planetary nebula.

It is an extremely dilute and hot environment where forbidden lines such as [O III] 5 007 - 4 959 Å appear very intense and narrow. Hydrogen and helium lines are always present.

We can see that almost all the flux is concentrated in the $H\alpha$ line and the two nitrogen lines [N II] on either side of it (the low-resolution spectrum does not separate them well) and on the two oxygen lines [O III] 5 007 - 4 959 Å.

During this very long and quieter phase, there is little change in the profiles of the lines present.

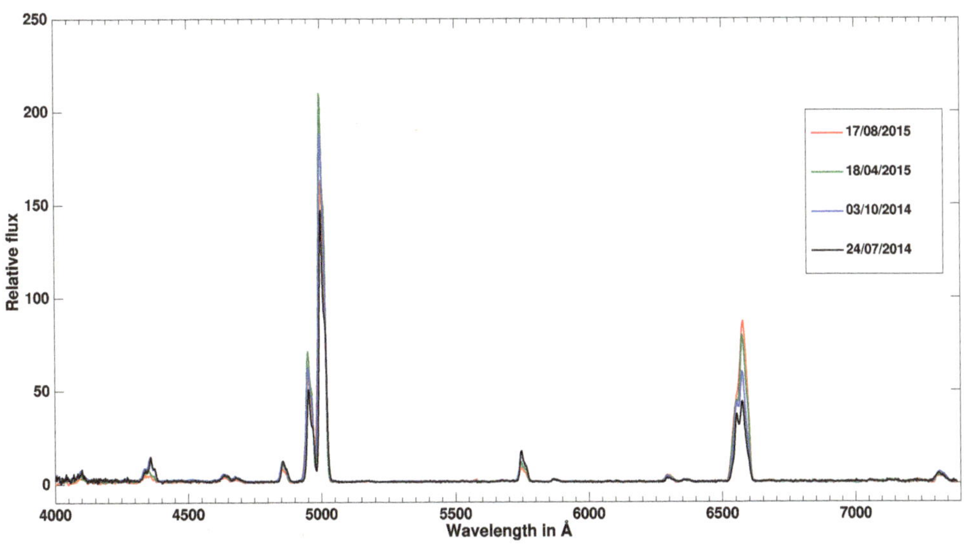

Fig. 6.7.5 — Nebular phase of V339 Del. 4 spectra made with a LISA spectrograph and an ATIK 414EX CCD camera on a C14 between July 2014 and August 2015. On the x-axis, wavelength in ångströms. On the y-axis, intensity relative to the continuum normalized to 1.

These spectra were produced at low resolution to reach magnitudes of 11 to 17 with a good signal-to-noise ratio, but with a loss of line detail.

Other novae appearing in the northern sky since and up to the writing of this book, were not all so easy to observe. Some because of their low luminosity or because they are often low on the horizon at less than 30 degrees above the ground for northern hemisphere observers (Scorpio or Sagittarius constellations). There were also a few novae in the Cassiopeia, Cepheus and Perseus constellations which didn't pose any observation problems, as they were quite high on the horizon (Nova Cas 2020, Nova Cas 2021, Nova Per 2018, Nova Per 2020, Nova Cep 2013, Nova Cep 2014).

6.8 Supernovae

Unlike novae, which are of galactic origin, the vast majority of supernovae are extragalactic. Known supernovae in the Milky Way are very rare. The literature mentions SN 1006 as the first galactic supernova observed, it appeared in the Lupus constellation in the year 1006. The Danish astronomer, Tycho Brahe (1546-1601), relates in his book *De Nova Stella*, the observation of a supernova on November 11, 1572 in the Cassiopeia constellation indicating that it is a “new star". Johannes Kepler (1571-1630) was able to observe one in 1604.

Another probably exploded around 1681 (Cassiopeia A) but there is no evidence of any direct observations at the time. The most famous is undoubtedly that of the year 1054, observed by Chinese astronomers for almost 2 years. It left in its place the remnant known as M1, the Crab nebula. Such a phenomenon could be seen in broad daylight due to its proximity and high luminosity, which could have exceeded an absolute magnitude of -19.

In 1885, SN 1885 was the first extragalactic supernova to be observed in M31 with an instrument. Since 1604, no supernova have been explode in our galaxy and while many stars are potential candidates, such as Eta Carinae, P Cygni, Betelgeuse, Antares and many others. Of course without, of course, knowing when they will explode in another mater: in 1 year, 10 years, 1 000 years, 10 000 years ?

The cases of recurrent novae RS Oph and U Sco are still under discussion: does the white dwarf gain mass and perhaps eventually explode as a supernova or lose mass in the nova explosions over time ?

Extragalactic supernova explosions are more frequent (there are many galaxies !) but with much lower apparent luminosities due to the distance of the host galaxies, spectroscopically accessible only at low resolution, or very low resolution for magnitudes greater than 16.

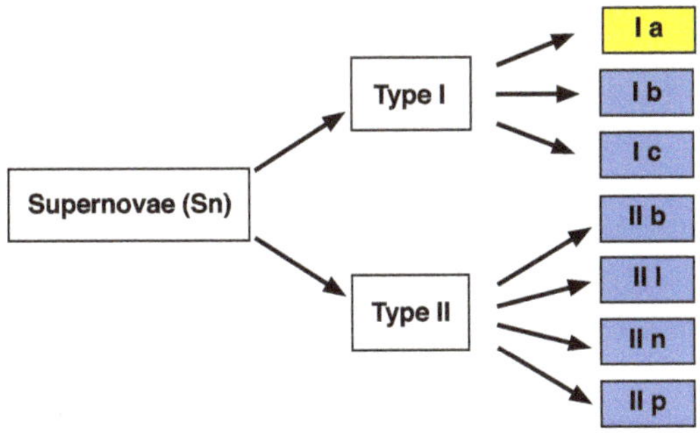

Fig. 6.8.1 — Main classification of supernovae.

According to their spectrum, the most widespread supernovae are classified into two main types, noted with Roman numerals, and into subclasses noted with letters:

- **Type I**: shows a spectrum without hydrogen lines.
- **Type II**: shows the presence of hydrogen lines in the spectrum.

Whatever the type of supernova, the spectrum shows very broad absorption lines, velocity signatures of several thousand km.s^{-1} with a strongly blue-shifted absorption component showing expanding matter coming towards the observer at velocity of the order of 10^4 km.s^{-1}. Of course, the redshift of the host galaxy must be taken into account (We therefore correct the spectrum by its value to measure the supernova's true expansion velocity).

Each type is subdivided into several subclasses. The following sections describe the most common subclasses:

6.8.1 Type Ia supernovae

This is the only subclass whose origin is the thermonuclear explosion of a white dwarf (degenerated star) having accumulated matter by accretion beyond the Chandrasekhar mass $M > 1{,}44\ M_{\odot}$ (solar mass). The star is destroyed. The absolute magnitude of SN Ia at maximum is always the same: absolute blue magnitude = -19.3 because it is always the same phenomenon. (actually this is a not strictly true. A correction has to be made based on the rate of decline in brightness). It is used to measure its distance or that of the host galaxy by comparing it with the maximum relative magnitude observed from Earth (Pogson's law). The temporal variation of light from this type of supernova is very regular and helps refine the classification.

The spectrum of a type Ia supernova is made up of silicon lines, but not helium.

The most recognizable signature is the presence of a broad silicon absorption line at 6 355 Å (at rest) but shifted dependent on the velocity of the explosion and the redshift of the host galaxy to typically around 5 800 to 6 150 Å.

We also note the presence of magnesium, iron, sulfur and calcium lines. When a Type Ia supernova reaches its nebular phase, you can also see lines from iron and atomically related elements.

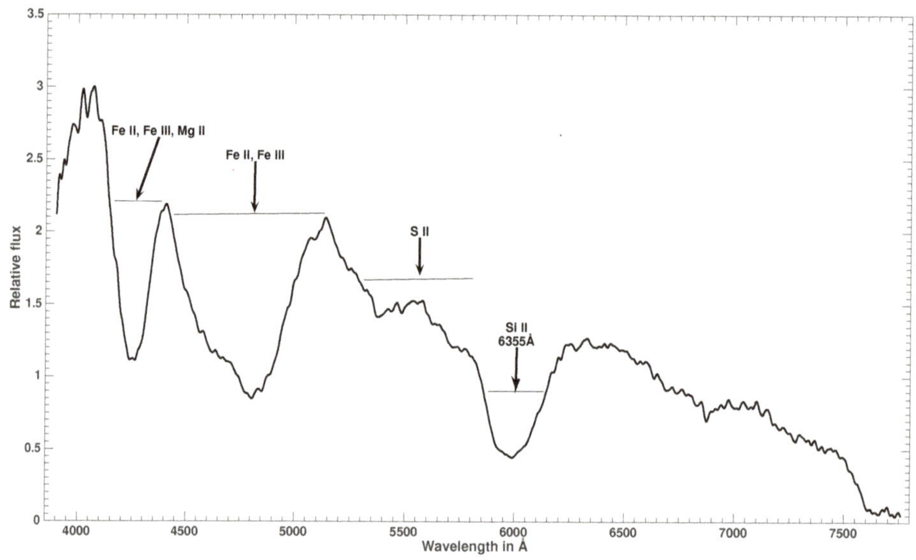

Fig. 6.8.2 — Spectrum of a Type Ia supernova. Spectrum of Sn 2021 aefx in the galaxy NGC 1566. Made with an Alpy 600 and an RC12 from Chile (10 x 1 200 seconds exposure time) on the 2SPOT installation. The lines that can be seen on this graph are iron (Fe II and Fe III), magnesium (Mg II), sulfur (S II) and silicon (Si II), enlarged by the expansion velocity and strongly blue-shifted. This spectrum is not corrected for the host galaxy's redshift z, which is 0.00473963 for NGC 1566 according to SIMBAD.

6.8.2 Other types of supernovae

Except for type Ia, all other types of supernova are produced by the collapse of the star's core. This applies to stars larger than 8 solar masses.

At the end of a massive star's life, fusion reactions of increasingly heavy elements occur. The star produces helium, carbon, neon, oxygen, silicium and the most stable element: iron.

Beyond that point, fusion reactions can no longer take place... there is no longer any radiation pressure: the star collapses towards its center under the effect of gravity: the iron core collapses in on itself, the peripheral layers fall and bounce off the degenerate core. This is the final explosion, spreading all these elements into space... (The short-lived radioactive elements produced by the supernova will quickly disappear).

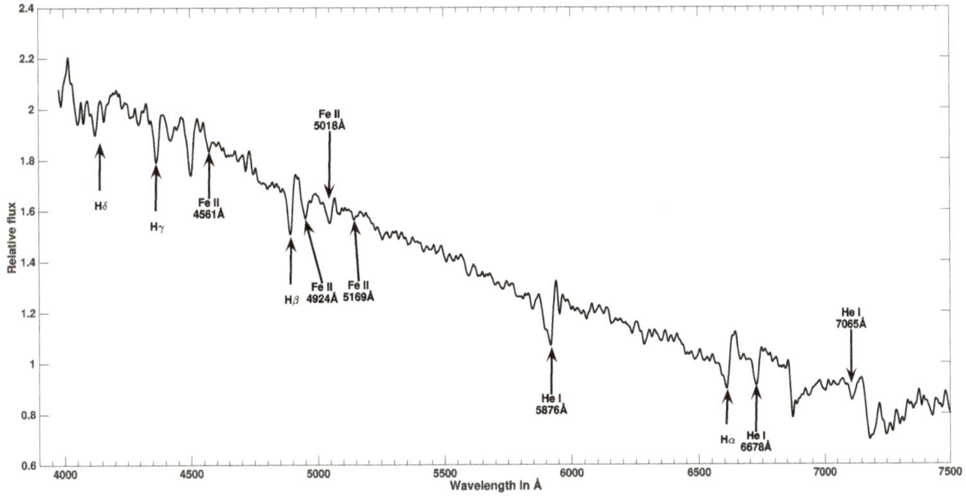

Fig. 6.8.3 — Spectrum of a type IIb supernova. Spectrum of Sn 2018 gjx in the galaxy NGC 865. Realized with a LISA and the T1m of the C2PU of the Côte d'Azur observatory on the Calern plateau (6 x 1 200 seconds exposure time). The spectral profile shows more intense hydrogen and helium lines, with a P Cyg-type profile.

These various reactions can be represented by successive layers of elements, with iron at the center, the term « onion peel » is often used to describe these dying stars...

Type Ib and Ic

The spectrum of a type Ib and Ic supernova shows helium lines but not silicon lines. Then in the nebular phase, we also note the significant presence of oxygen lines. The difference between the Ib and Ic subclasses is determined by the abundance of helium. In the Ic class, helium is less abundant than in the Ib class.

Type IIb

The spectrum of a type IIb supernova shows helium lines more intense than hydrogen, with a P Cyg profile. Otherwise, it is type II or type "normal".

Other subclasses

There are many other sub-classes, such as:

• The **type II l**: magnitude decay is linear, the letter "l" standing for "linear".

• The **type II p**: where the letter "p" denotes a "plateau" in the decay of magnitude over time.

According to the Transient Name Server website, which has compiled statistics on supernova types, there have been identified (between January 1st, 2016 and March 2024):

• 9 086 Sn of type Ia, or almost 66%.
• 2 068 Sn of type II, or 15 %
• 406 Sn of type IIn, or 2,9 %
• 294 Sn of type Ic, or 2,1 %
• 250 Sn of type Ib, or 1,8 %
• 231 Sn of type IIp, or 1,7 %
• 196 Sn of type IIb, or 1,4 %

Subclasses not described in this chapter account for less than 10% of the total. All statistics can be consulted by following this link:

https://www.wis-tns.org/stats-maps

We can see that extragalactic supernovae are not rare phenomena (there are many galaxies, and supernovae are very luminous!!) because in 8 years, there have been over 13 000 confirmed supernovae, the vast majority which are type Ia.

6.9 Supernova spectrum acquisition

As extragalactic supernovae are often of low apparent luminosity, you need a low-resolution spectrograph like the Star'Ex LR, UVEX or LHIRES III (equipped with a

300 lines/mm grating). The Alpy 600 and LISA are also very well suited to this task and will give excellent results, it is with these 2 spectrographs that the observations in this chapter were made. Depending on the diameter of the telescope, magnitudes of 15 and even 16 can be reached, Beyond that, it may be worth using an even less resolving grating of 150 or 200 lines/mm. Resolution will certainly be degraded around $R = 130$, but for the classification of a supernova it is hardly a problem because the characteristic lines are very wide and easily identifiable.

While very low resolution can be used and we can then attempt targets of magnitude 18 to 19, you'll need a very dark sky to be able to produce the spectra, because if the magnitude of the background sky is lower, you will not be able to see the coveted object ! The Alpy 200 was designed for this purpose. It was originally developed by English amateur Robin Leadbeater who used it with a 280 mm aperture Schmidt-Cassegrain telescope to confirm and officially classify numerous supernovae, including some with a magnitude $V = 18.6$, and to measure quasars with redshifts up to $z = 4.5$.

All his work can be seen on his website:

http://www.threehillsobservatory.co.uk/astro/astro.htm

Supernovae may also appear in galaxies close to our own, such as M31, the large or small Magellanic cloud. In these cases, we will probably be able to use a more resolving spectrograph, as the target will be brighter.

6.10 Example with SN 2021 pit

Let us take the example of spectra made on supernova SN 2021 pit which exploded in the galaxy NGC 1448 (discovered on June 10, 2021).

SN 2021 pit's coordinates are:

RA: 03h 44' 30,21" (J2000)
DEC @: -44° 37' 57,25" (J2000)

It was therefore only visible from the southern hemisphere. We were able to make spectra with our 2SPOT team remotely from Chile (Alpy 600 spectrograph on a 305 mm diameter Ritchey-Chrétien telescope and an ATIK 414EX CCD camera used in 1x1 binning). The supernova had a magnitude $V = 13.5$ at the time we were able to produce the first spectrum on June 12, 2021.

Fig. 6.10.1 — Position of SN 2021 pit in NGC 1448. On the left, the DSS-2 survey reference field, on the right the image taken with the autoguider camera showing the position of the supernova (ASI 174 mini CMOS camera: 30 seconds exposure).

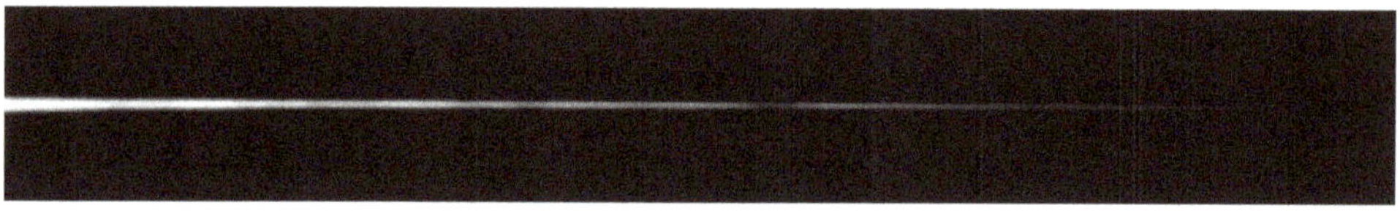

Fig. 6.10.2 — Raw 2D spectrum of SN 2021 pit. Sum of 6 spectra of 600 second each obtained with an Alpy 600 equipped with a 23 μm slit.

We were then able to carry out further spectra on June 13, 19, 20 and 21, 2021 with the same total exposure time of one hour (6 x 600 seconds). As for any acquisition, the supernova must be well centered in the spectrograph slit and autoguide on a nearby star, brighter if possible. The presence of other, brighter stars in the sensor's field of view enabled precise autoguiding, while ensuring that the supernova did not emerge from the slit, which would result in a loss of flux and therefore a poorer signal-to-noise ratio.

Looking at the line-free spectral profile of hydrogen, we can estimate that this is type I. It shows silicon but no helium lines, so it is probably a type Ia supernova. For confirmation of the profile obtained, you can use a handy online tool: the Gelato website from the Canary Islands Astrophysics Institute:

https://gelato.tng.iac.es/

First, you need to register by entering your username and password. Next, you have access to the interface shown in Figure 6.10.4.

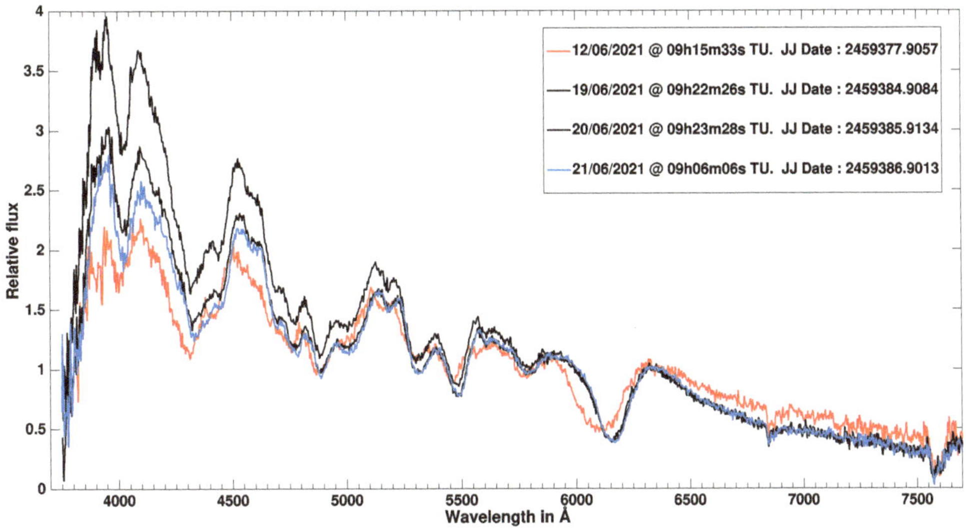

Fig. 6.10.3 — Sn 2021 pit spectrum evolution. 4 spectra observed between June 12 and 21, 2021. The continuum of the 4 spectra was normalized to 1 at around 5 700 Å. On the x-axis, wavelength in ångströms. On the y-axis, intensity relative to the continuum normalized to 1.

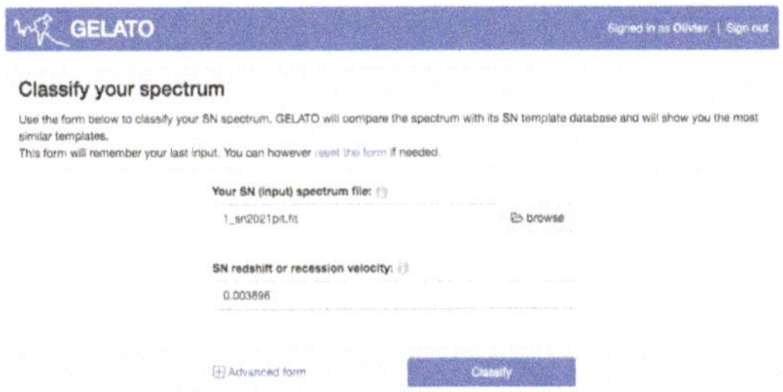

Fig. 6.10.4 — Gelato interface. To automatically classify the spectrum of a supernova.

Upload your spectrum in .fit or .fits format: enter the file name in the first box and the redshift of the host galaxy in the second (search for redshift in SIMBAD). Here, the galaxy is NGC 1448 with $z = 0.003896$. Then validate by pressing "**Classify**". The result will be displayed after a few seconds.

Gelato's answer indicates a type Ia supernova and gives a confidence index, here 100 % (see Figure 6.10.5).

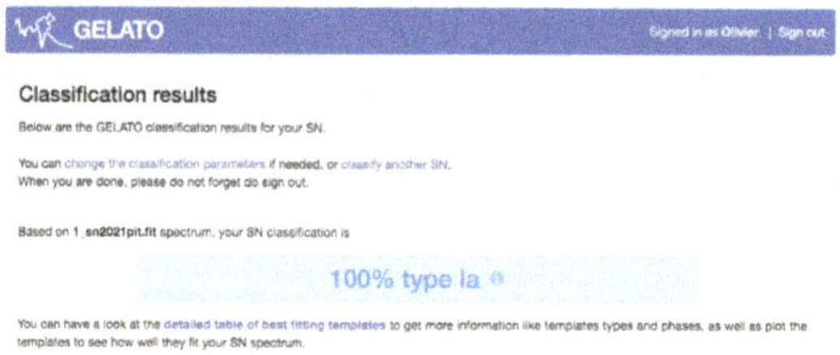

Fig. 6.10.5 — Classification result by Gelato.

6.11 Calculating the velocity of the supernova ejecta

In the case of a Type Ia supernova, we can calculate the speed of the ejecta by taking, for example, the silicon line, which is normally at rest at 6 355 Å. It is well isolated from the others, allowing precise measurement of its shift.

If you have a series of spectra taken over several days, we can measure the variation in expansion velocity. If the supernova is of a type other than Ia, we can use hydrogen or helium lines.

As the spectra produced include the Earth's velocity through space, to achieve greater precision we could correct for heliocentric velocity. We will start by correcting each spectrum.

We could almost omit this step, since the velocity we will be measuring is much higher than this correction which will induce a maximum error of ± 30 km.s^{-1}. We only mention it here as a reminder that the spectra we produce include the Earth's velocity.

Then we measure the wavelength value of the Si II line 6 355 Å, shifted towards blue. To accurately determine this value, you can use the various methods described in the preceding chapters such as modeling a Gaussian curve on the line to be measured. Tools for this purpose exist in various software packages such as Visual Spec or ISIS.

We then measure the spectral shift between the value of the line at rest and the measured value, then calculate the velocity according to the Doppler-Fizeau equation we've already encountered in the paragraph on calculating galaxy redshift:

$$V_r = \frac{\lambda_{Obs} - \lambda_0}{\lambda_0} c$$

Where:
V_r: the radial velocity of the ejecta in km.s^{-1}.
λ_{Obs}: the measured value of the Si II line.
λ_0: the resting value of the Si II line, i.e. 6 355 Å.
c: the speed of light: 299 792.458 km.s^{-1}.

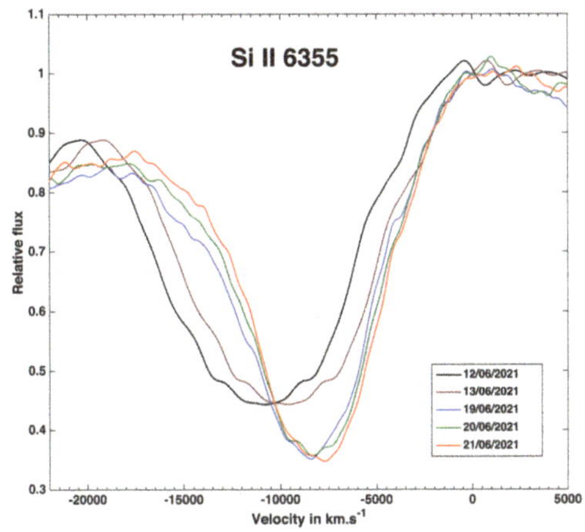

Fig. 6.11.1 — SN 2021 pit velocity graph. Carried out between June 12 and June 21, on the silicon line at 6 355 Å. For greater clarity, the curves have been filtered. On the x-axis, the velocity in km.s^{-1}. On the y-axis, the relative intensity with respect to the continuum normalized to 1 at the wavelength of the silicon line at rest.

Date	V Helio	Si II λObs	Si II λo	Vr Obs	z NGC1448	JJ-2459000	Vr cor abs
12/06/21	10.569	6125,81	6355	-10812	0,003896	377,9057	**11980**
13/06/21	10,726	6150,35	6355	-9654	0,003896	378,9121	**10822**
19/06/21	12,689	6175,97	6355	-8446	0,003896	384,9084	**9614**
20/06/21	11,728	6178,01	6355	-8349	0,003896	385,9134	**9517**
21/06/21	11,837	6191,06	6355	-7734	0,003896	386,9013	**8902**

Fig. 6.11.2 — Table of velocity measures on SN 2021 pit. Synthesis of the various measurements and calculations to arrive at the ejecta velocity value in the last column on the right.

You can then generate a velocity variation curve; the more measurement points, the better the accuracy. The velocity measured here includes the redshift of the host galaxy (NGC 1448) which must be subtracted to determine the ejecta's true velocity. The redshift value of NGC 1448 is $z = 0.003896$, which means a velocity of 1 168 $km.s^{-1}$.

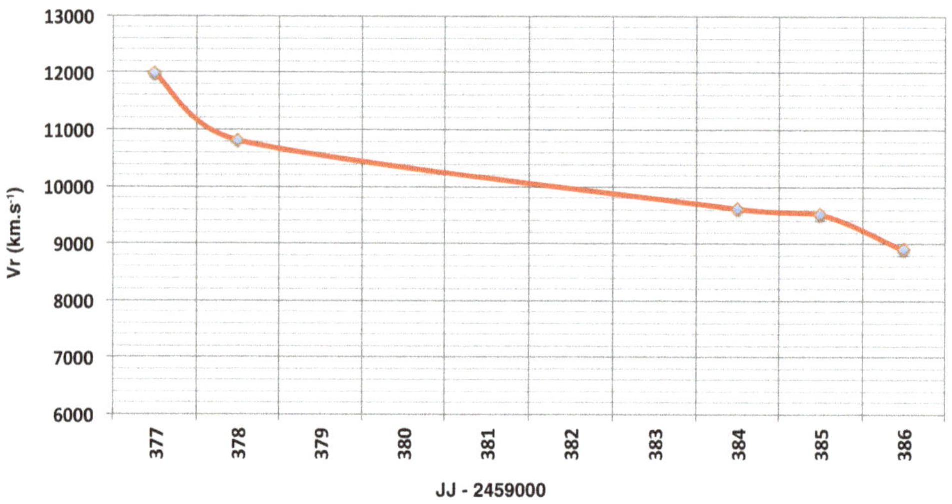

Fig. 6.11.3 — Corrected velocity variation curve on SN 2021 pit. The curve shows the deceleration of the ejecta velocity between June 12 and 21, 2021, corrected for the redshift velocity of the host galaxy. On the x-axis, time in Julian day. On the y-axis, velocity in $km.s^{-1}$.

The curve shown in Figure 6.11.3 was carried out with just 5 measurement points over a 9-day period. There is a wide gap between the June 12 and 13 measurements and those of June 19, when the variations become more significant.

We can see that it is imperative to carry out spectra as soon as possible after the supernova has been discovered and produce at least one spectrum per night for the first few days. After a week, variations in expansion velocity from one day to the next are less significant. In the case of this supernova, the first measurement was taken 2 days after its discovery on June 10, the velocity measured on June 12 was probably not the highest. Decrease is most rapid in the first few days but tends to slow down thereafter.

6.12 Bibliography

- Classical Novae, second edition by M.F. Bode and A. Evans, (Cambridge Astrophysics Series, 2008). ISBN 978-1-107-40534-9.

- Spectroscopy of Novae – A User's Manual by Steven N. Shore (https://doi.org/10.48550/arXiv.1211.3176)

- Stars and their spectra second edition, An introduction to the spectral sequence by James B. Kaler, (Cambridge University Press, 2011). ISBN 978-0-521-89954-3.

- Classical and recurrent novae by Ulisse Munari, The Journal of the American Association of Variable Star Observers, vol. 40, no. 1, p. 582, juin 2012.

- Spectral modelling of Type IIb Supernovae. Comparison to SN 2011dh and the effect of macroscopic mixing by Mattias Ergon and Claes Fransson, Astronomy and Astrophysics, juin 2022, A&A 666, A104, 2022.

- Atlas of photometric and spectroscopic daily monitoring of the very slow Nova Cas 2021 (=V1405 Cas) during the first 660 days of its outburst by P. Valisa, U. Munari, S. Dallaporta, A. Maitan, and A. Vagnozzi (https://doi.org/10.48550/arXiv.2302.04656).

- Spectroscopie des Supernovae de type Ia des expériences SuperNova Legacy Survey et Nearby SuperNova Factory pour la cosmologie by Flora Cellier-Holzem (thesis defended october 4, 2013).

- SN 2019ein: New Insights into the Similarities and Diversity among High-velocity Type Ia Supernovae by Miho Kawabata et al., The Astrophysical Journal, 893:143 (16pp), 2020 April 20.

- Type IIb supernova Sn 2011 dh, spectra and photometry from the ultraviolet to the near-infrared by G. H. Marion et al., The Astrophysical Journal, 781:69 (18pp), 2014 February 1, doi:10.1088/0004-637X/781/2/69.

- Multiwavelength Modeling of the SED of Nova V339 Del: Stopping the Wind and Longlasting Super-Eddington Luminosity with Dust Emission by Augustin Skopal, The Astrophysical Journal, 878: 28 (18pp), 2019 June 10, https://doi.org/10.3847/1538-4357/ab1f07.

- Les novae by François Teyssier on his website: http://www.astronomie-amateur.fr

- A Speed Bump: SN 2021aefx Shows that Doppler Shift Alone Can Explain Early Excess Blue Flux in Some Type Ia Supernovae by C. Ashall *et al.*, The Astrophysical Journal Letters, 932: L2 (8pp), 2022 June 10, https://doi.org/10.3847/2041-8213/ac7235.

- Supernovae by Paul Murdin and Lesley Murdin, (Cambridge University Press, 1978). ISBN 0-521-30038-X.

- Optical Spectra of Supernova by Filippenko, Alexei V. (Annual Review of Astronomy and Astrophysics, Volume 35, 1997, pp. 309-355). DOI: 10.1146/annurev.astro.35.1.309

7. Confirming a planetary nebula

The purpose of this chapter is to show you how, using amateur techniques, to determine the nature of an object that is referenced but not classified in an astronomical catalog.

Over the past decade, many amateur astronomers have discovered, from their own images or with the help of professional observation programs, new non-referenced objects whose nature is yet to be determined. Because a spectral signature removes any doubt about the true nature of an object, this is one of the few fields in which classical astronomical imaging and spectroscopic observations blend and complement each other.

"Classical" imaging essentially uses narrow-band filters of a few nanometres passband, centred on the nebular emission line wavelength of elements such as sulphur, hydrogen or oxygen (SHO).

Low-resolution spectroscopy allows the light to be "broken down" into a spectrum that can be physically measured in the visible range between 400 and 750 nm and also in the near ultraviolet range. An atypical case of the latter is described at the end of this chapter.

Investigations by amateurs have led to the discovery of several hundred new objects, some very small, others more extensive, but all of them with very faint magnitudes. They are listed and initialled with the discoverer's name. Each is classified at this stage of investigation as "Candidate planetary nebula". One of theses objects is selected for confirmation because it shows the morphology of a planetary nebula through a $H\alpha$ and/or [O III] filter. To confirm the identity, a spectrum is required, which is not always easy to obtain!

This chapter offers a few tips to explain the process and maximise the chances of success.

7.1 The story of the LDu 1 planetary nebula

Many amateur astronomers have discovered new objects in their own images. Nicolas Outters was the first amateur astronomer to make discoveries with his deep-sky images. He was followed by other amateurs such as Pascal Le Dû, Lionel Mulato, the APO Team in Chile (Thierry Demange, Richard Galli and Thomas Petit), being the most productive ones. We are going to tell you all about the story of the LDu 1 planetary nebula from its discovery by Pascal Le Dû in 2011 on an image to its confirmation by spectrography in 2016.

In 2011, as he photographed the Sharpless SH2-124 catalog nebula from the tip of Brittany, Pascal noticed something on the $H\alpha$ layer at the top right of his image: a small circular spot, 2 arcminutes in diameter, not identified in astronomical catalogs. Intrigued, he contacted astrophysicist Agnès Acker, a specialist in planetary nebulae.

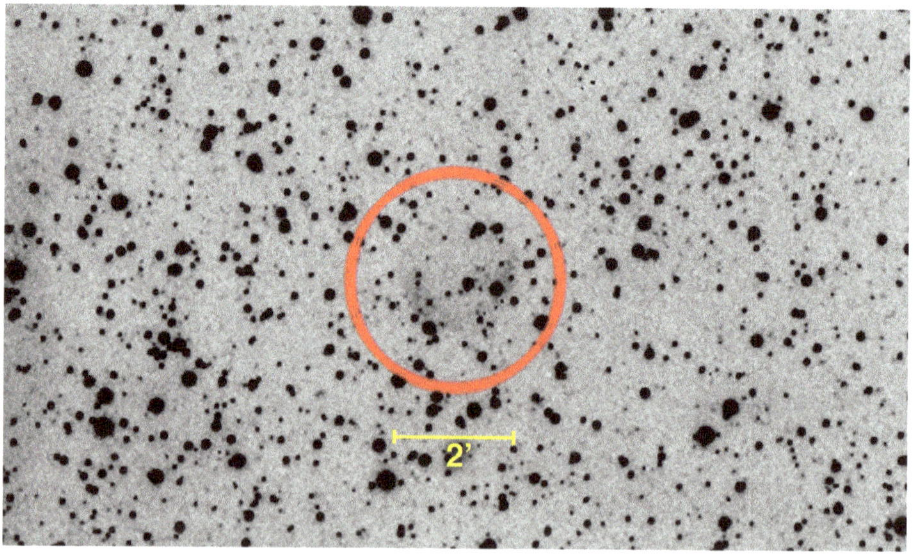

Fig. 7.1.1 — Detail of LDu 1 nebula image. Part of the image of SH2-124 taken by Pascal Le Dû with a $H\alpha$ filter attached on a Takahashi FSQ106EDX refractor equipped with a QSI583 CCD camera on a Losmandy G11 mount. 400 minutes of exposure on this layer.

On August 23, 2011, Pascal received this reply from the astrophysicist:

"Hello,
Bravo, your object is a new LDu 1 planetary nebula!
Confirmation with these images in r, i and H alpha (do not pass them around).
But a spectrum will be necessary as usual.
Congratulations!
Agnès Acker"

Then a few years passed and in 2016, Pascal Le Dû, Thierry Lemoult and Olivier Garde, submitted a mission request to confirm planetary nebulae using spectroscopy at the Saint-Véran observatory in the French Alps (run by the Astroqueyras association).

The aim of this mission is to produce spectra of planetary nebulae candidates at very low magnitudes, under a dark sky, far from any light pollution, using a 500 mm diameter Ritchey-Chrétien telescope with a focal length of 4 000 mm. This instrument being equipped with a low-resolution spectrograph, the LISA (R = 1 000).

The mission was accepted by the Astroqueyras program committee which allocated week S35 (from August 27 to September 4, 2016).

Arriving on site, we installed the LISA spectrograph on the telescope, combined with an ATIK 414EX CCD camera to acquire the spectra. The 6.45 μm pixels provided good sensitivity and a fairly low reading noise of around 5e^-, although this is greater than that of current CMOS technology cameras.

For the autoguider camera, we chose an ATIK 314L+, which had a fairly large sensor and thus offered a fairly wide field of view, an important detail due to the RC500's high focal length. The camera's pixel size was also 6.45 μm, which provided good detection for optimal positioning of the target in the spectrograph slit.

The mount used was an AP1600 from Astrophysics equipped with absolute encoders on the two RA and DEC axes. These allowed for excellent target pointing accuracy, a good target tracking and highly accurate autoguiding.

Fig. 7.1.2 — The planetary nebula candidate confirmation team. With (from left to right) Olivier Garde, Pascal Le Dû, Thierry Lemoult. In the background, the RC 500 at the Astroqueyras observatory, equipped with the LISA spectrograph.

Here are the important points to consider when producing quality spectra of planetary nabulae:

• Observing under the darkest possible sky, free of light pollution, if possible at altitude, to reduce interference from the effects of the Earth's atmosphere. The Saint-Véran observatory is located at an altitude of 2 990 m. Moonless nights are preferable.

• Use a high-performance, low-resolution spectrograph such as the LISA, the Star'Ex LR or the Alpy 600. The UVEX or the LHIRES III are not recommended for this type of object for the following reasons: The LHIRES III has been optimized for F/10 ratio and high-resolution use, it can be used with lower resolution gratings but its optical configuration is not very well suited to producing spectra of very faint objects.

Regarding the UVEX, its Czerny-Turner optical design known as "crossed" is not well suited to producing spectra of extended objects. You can also use it if the candidate is a one-off object and/or is of interest in the near ultraviolet range, as you will discover at the end of this chapter with candidate Pa 30 (see section 7.3.5).

• Bring the telescope's focal ratio as close as possible to F/5 if using a LISA, or even F/4 with an Alpy 600. In the case of the RC 500 (F/8 natively), we used a CCDT67 focal reducer from Astrophysics. This reducer is well documented to determine precisely the right back focus to adopt and thus obtain the desired reduction ratio (the distance between the focal reducer and the spectrograph slit).

• On telescopes with focal lengths over 2 000 mm, the slit width must be increased to 35 μm, or even 50 μm to limit light flux losses. We used a 50 μm slit with the RC 500, whose focal length had been reduced to 3 000 mm.

• Ensure excellent tracking during autoguiding, which means no loss of flux to the target and hold it continuously in the spectrograph slit.

• Achieve unit exposures from 900 to 1 200 seconds and do at least 1 to 2 hours of it to improve the signal-to-noise ratio of the spectrum. Exposure time adjustment can be optimized after the first image, depending on the intensity of the spectrum.

It was on the night of August 26 that we were able to obtain the spectrum of LDu1. We pointed the target at its center coordinates:

RA: 21h 36' 05,5" (J2000)
DEC: +50° 54' 10" (J2000)

It is useful to center the target in the spectrograph slit. Nebula is not point source, but it covers an area of about 2 arc minutes. We positioned the slit on the most contrasting part of the nebula, to the south, clearly visible on Pascal's $H\alpha$ image from 2011 (see Figure 7.1.3).

On an exposure, even a long one with the autoguiding sensor, the object was not visible from the background, but we recognized the field with the stars in its environment, which allowed target pointing to be fine tuned.

Figure 7.1.3 shows how the slit was positioned over the most intense area of the nebula. The slit appears as a bright line in the image on the left. The yellow arrows indicate the area of interest and centering of the slit. The height of the target above the horizon was checked before acquisition began:

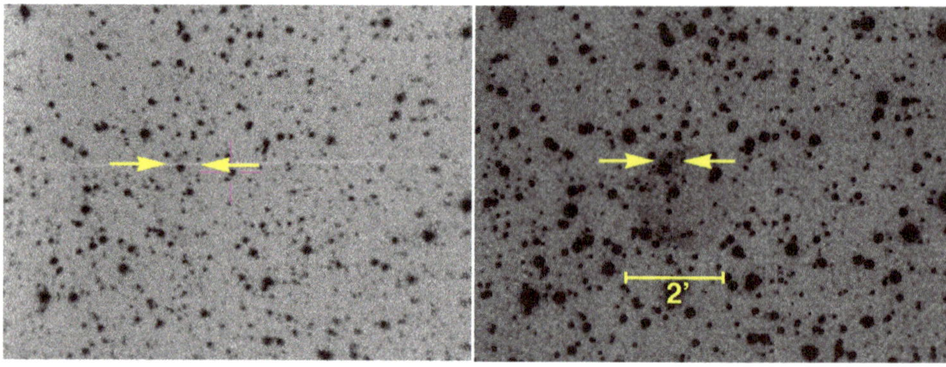

Fig. 7.1.3 — Position of spectrograph slit on LDu 1. On the left: image of the autoguider sensor, on the right: $H\alpha$ image created by Pascal in 2011. This negative 2011 image helps us define the centering zone for the slit (yellow arrows).

• Note: if the target is below 30° of altitude, the atmospheric layer crossed by the light coming to the telescope is greater than two air masses (twice the thickness of the Earth's atmosphere at the zenith), which will reduces and reddens the already weak flux. It is even more so if the target is lower.

• The target must be tracked for 3 hours without the mount having to cross the meridian and execute a meridian flip. In case of meridian flip, it could make spectrum processing more tedious due to reversal of the direction of the slit on the sky.

Then start the autoguiding on a star close to the slit. The exposure time on the selected star has been determined so as not to saturate it. If you have a good mount, you can space out the time between each exposure.

Autoguiding with a short exposure time, less than 1 second, for example, is not recommended as guidance may chase the astronomical seeing, making the object oscilate between the slit and outside, which results in the loss of valuable flux to the spectrograph.

On LDu 1, a series of 9 exposures of 1 200 seconds was performed, for a total of 3 hours. The signal remained very weak on each spectrum, close to noise. As usual, a series of FLAT lamp exposures is carried out, around thirty with an ADU level close to 50 000 ($\approx$ 75 % of the sensor dynamic range) to optimize the signal in the blue part of the spectrum. A reference star was also observed at the same altitude as the target.

Lastly, spectra of the calibration lamp were taken just after the last exposure and without moving the telescope position.

Fig. 7.1.4 — 2D spectrum of LDu 1. Stack of 9 exposures of 1 200 seconds and identification of the most obvious lines.

On the sum of all raw spectra of LDu 1, there did not get much signal, but we could still distinguish the main emission lines with the $H\alpha$ line surrounded by the 2 [N II] lines and the [O III] line 5 007 Å. Other lines are also visible, but well hidden in the noise.

Spectrum processing needed to be meticulous. To bring out the object's signal, which is often very weak, we carefully adjusted the display thresholds. It was essential that the spectrum binning zone and the sky background subtraction zone be clearly defined and must not contain any trace of the nebula, otherwise the object's faint signal would be suppressed.

Several tests were performed by changing the width of the binning zone and the sky background subtraction zones each time. This can be a long and tedious process. But the aim here was to produce the cleanest possible spectrum, which highlighting the desired signal.

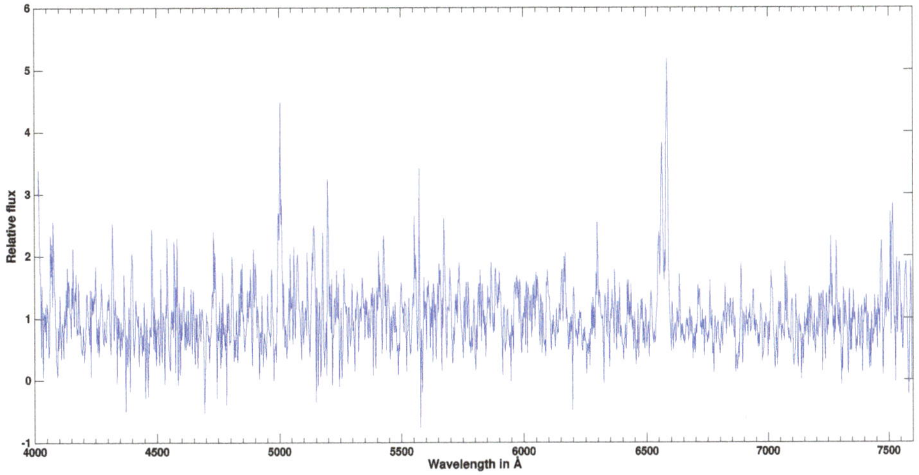

Fig. 7.1.5 — 1D spectrum of LDu 1. The result of the treatment shows the presence of [OIII], [N II] and $H\alpha$ lines. The signal-to-noise ratio of the spectrum obtained is low, but sufficient to identify the main emission lines.

The final spectrum remained fairly noisy (see Figure 7.1.5) which is somewhat normal in this case, as the target was very faint. But some emission lines could be identified. On this type of object, almost all the flux is concentrated on a few emission lines, which is why we can achieve a positive result.

The spectrum was them submitted to Agnès Acker so that she could check wheter it was consistent with that of a planetary nebula.

The presence of the [O III] and $H\alpha$ lines was a good criterion for retaining the object, but further analysis was carried out by Professor A. Acker, such as the intensity ratio of certain lines to each other.

Some time later, we received the confirmation that LDu 1 had been classified as a planetary nebula.

Query : PN Ldu 1

Basic data :

NAME LDu 1 -- Planetary Nebula

Other object types:	**PN** (2013MNRAS,PN)
ICRS coord. *(ep=J2000) :*	**21 36 05.5 +50 54 10 [ ] D** 2012RMxAA..48..223A
FK4 coord. *(ep=B1950 eq=1950) :*	21 34 20.1 +50 40 41 [ ]
Gal coord. *(ep=J2000) :*	094.5840 -00.8909 [ ]

Fig. 7.1.6 — The LDu 1 object referenced in SIMBAD. By making a request "PN LDu 1" in the database.

This can be checked in the SIMBAD database:

https://simbad.cds.unistra.fr/simbad/

This confirmation was not the easiest to achieve, because the nebula is very dim and extensive. As a reminder, it took Pascal 400 minutes of imaging $H\alpha$ to detect his signal. It is important to choose the most intense zone of nebulosity, which is not always easy especially if you do not have a well-exposed image that details all the contrasting areas of the object.

In this respect, some "attractive" colour images of theses type of objects show highly contrasting objects that have only been revealed through extensive processing, by applying high coefficients to the weakest signals. It is not feasible to achieve the same result with spectroscopy. The signal on the observed object is not amplified, but returned as it is.

7.2 How to confirm a planetary nebula?

At the time of writing this book, the list of candidate planetary nebulae contains over 1 200 objects. Just over 500 have a spectrum and 150 are identified as true planetary nebulae. There are still many targets without spectra to be confirmed. The list can be found on the website:

https://planetarynebulae.net

This website is managed by two amateur astronomers: Pascal Le Dû and Thomas Petit. This is where you'll find your future targets, get information on how to observe them, as well as tips on how to submit your observations.

To help you get started, start with a candidate which already has a spectrum. This will enable you to check whether your entire acquisition and processing chain gives a similar result.

It is important to validate your methodology and equipment before attempting a candidate that has never been observed before. We advise you to start your adventure with bright, pointlike targets that have a good planetary nebula morphology. Next, you can try out weaker, more diffuse targets by carefully studying the available images centring the slit on the most fruitful areas.

Once you have achieved the spectrum of a candidate, you can send your comments to Pascal Le Dû, who centralizes worldwide candidate discoveries and spectral observations. The message to Pascal must contain the following information:

• Your first name and surname
• The object's name and coordinates in J2000
• Spectra in fit or fits and ASCII format (.dat), and if possible, the .log and .xlm files such as those generated by ISIS.
• The spectroscopic observation sheet, completed as fully as possible, in pdf format, downloadable from this link (all explained):

https://planetarynebulae.net/documentation/ExempleFicheSpectro.doc

• If possible, an image of your autoguider sensor showing the target centered on the spectrograph slit.

• Your comments (place of observation, hardware configuration used, reasons why this signal was selected...).

Then send it to Pascal at his e-mail address:

pascal.ledu@2spot.org

This information will then be sent to Quentin Parker, from the University of Hong Kong. This astrophysicist is now in charge of officially confirming the candidates as Agnès Acker is now retired. Your spectrum can be included in the HASH database (The University of Hong Kong/Australian Astronomical Observatory/Strasbourg Observatory H-alpha Planetary Nebula) which brings together all the planetary nebulae discovered in this way:

http://202.189.117.101:8999/gpne/index.php

Your data may be used in the context of a future article or publication (and you may be cited as a co-author).

For some candidates, it can be very difficult to make an observation that will enable the object to be classified. However, very faint magnitude spectra can be produced when the object is a planetary nebula, because most of the flux is concentrated only on certain emission lines.

It is this concentration of flux on a few lines that allows us to confirm the identity of a planetary nebula. Indeed, this type of object rarely shows the continuum spectrum of the central star, because the star is so faint. Only the nebula's emission lines stand out from the noise.

Do not be discouraged if your first spectra are unsuccessful. After a few unsuccessful attempts and with the experience acquired, you will quickly obtain spectra that can be analyzed.

7.3 Some examples of the spectra obtained

The spectrum of a "candidate" planetary nebula selected thanks to its imaging morphology, can reveal an entirely different object when its spectrum is analyzed. This may involve, for example of a dwarf nova, a Be star, of H II zones, a galaxy, a symbiotic star, a supernovae remnant, etc. Hence the importance of obtaining a spectrum, to determine the nature of the object.

In the following section, we are going to show you some spectra acquired during our various planetary nebula confirmation campaigns with Stéphane Charbonnel, Pascal Le Dû, Thierry Lemoult, Lionel Mulato and Thomas Petit.

7.3.1 Pre 24

Object discovered by Norwegian astronomer Trygve Prestgard who studied at the University of Grenoble in France, where he defended his thesis on "the spectroscopy of minor bodies", and also identified more than 70 candidates on various surveys such as SDSS et les images Wise. Pre 24 has the following coordinates:

RA: 04h 25’ 53,60" (J2000)
DEC: +39° 49’ 10,4” (J2000)

We estimated the object’s magnitude at $V = 20$ to 21, by comparison with surrounding stars. But it remains difficult to estimate for extended objects.

During a mission to the C2PU on the Calern plateau in September 2017, we produced a spectrum of the Pre 24 object using a 1 m diameter Cassegrain telescope and a LISA spectrograph equipped with a 50 µm slit reducing the resolution from $R = 1\ 000$ to about $R = 500$.

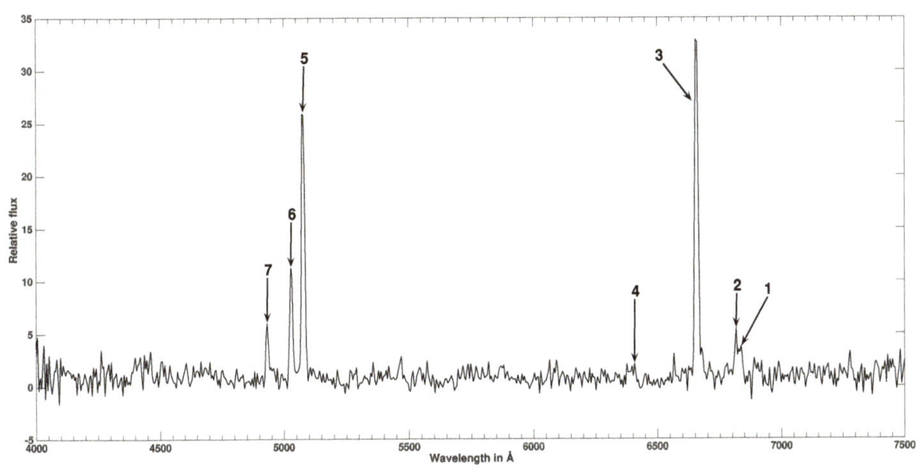

Fig. 7.3.1 — Pre 24 1D spectrum. Performed with a LISA spectrograph equipped with a 50 μm slit on a 1 m diameter Cassegrain telescope and an ATIK 414EX camera: 4 x 900 seconds exposure. On the x-axis, wavelength in ångströms. On the y-axis, intensity relative to the continuum normalized to 1. The numbers indicate the emission lines described in the table of figure 7.3.2.

We used a focal reducer Astrophysics CCDT67 to bring the telescope's F/D ratio close to F/6. The first 900-second exposure on this target showed us several lines in emission and we accumulated 4 exposures for a total of 1 hour time of exposure.

The spectrum shown in Figure 7.3.1 strongly look like that of a planetary nebula, with the presence of the $H\alpha$, [S II] and [O III] emission lines, but if you check the position of these lines, you'll see that they're all red-shifted.

It is an extragalactic object, a galaxy with an active nucleus, and its redshift (i.e. distance) can be measured using the visible lines.

The question arises as to how to capture a spectrum of this very faint target at around magnitude 20-21.

Of course, the telescope's diameter was a great help, but we also found that were dealing with very intense lines, some at more than 10 or even 30 times the continuum, which made them stand out from the noise and therefore observable.

Almost all the flux was concentrated on just 4 of the main emission lines.

Elements	line #	λ_0	λ_1	$\Delta\lambda$	Rel. Flux	z
[SII] 6731	1	6730,85	6826,62	95,77	3,45	0,01423
[SII] 6716	2	6716,47	6816,32	99,85	5,53	0,01487
Hα	3	6562,82	6658,00	95,18	5,88	0,01450
[SIII] 6312	4	6312,10	6407,08	94,98	2,16	0,01505
[OIII] 5007	5	5006,84	5078,92	72,08	11,27	0,01440
[OIII] 4959	6	4958,92	5030,43	71,51	25,87	0,01442
Hβ	7	4861,33	4931,91	70,58	32,80	0,01452
					Average z =	**0,014569**

Fig. 7.3.2 — Calculating the redshift of Pre 24. Using the 7 lines identified in figure 7.3.1. Values of the column "Rel. Flux" correspond to the intensity measured by modeling a Gaussian curve on the profile of each emission line.

We measured an average redshift $z = 0.01457 \pm 0.00029$, a receding velocity in the order of 4 360 km.s^{-1}, representing a distance of 59 Mpc (approximately 192 million light-years) using a Hubble constant value of H_0 =74.03 km.s^{-1}.Mpc^{-1}.

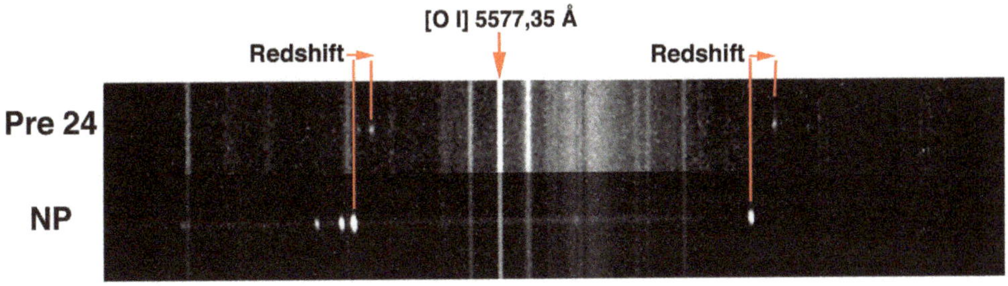

Fig. 7.3.3 — Pre 24 2D spectrum. The comparison of the spectrum of a classic planetary nebula (PN, Kn 66 here) with the spectrum obtained from Pre 24, shows a red-shift in the spectrum. The [O I] airglow line and other atmospheric or light pollution lines... allow us to verify that this shift is real.

This example shows that it is important to check the position of the various lines in a spectrum before hastily concluding it to be that of a planetary nebula. At first glance, the spectrum of an active-core galaxy, such as a Seyfert type 2 galaxy, is quite similar to that of a planetary nebula on first approach. So be careful and don't rush into anything!

The shift can be real, as was the case with this candidate, but it can also be a spectral calibration error and it will be necessary to determine the reason for this discrepancy. We can easily check that it is real by processing the spectrum without subtracting the background sky in order to visualize the auroral or airglow line [O I] 5 577.35 Å which can be seen on exposure times of over 600 seconds and must be at the correct wavelength.

7.3.2 Ra 69

Discovered by French amateur astronomer, Thierry Raffaelli, who has discovered more than 80 candidates on various surveys. Ra 69 has the following coordinates:

RA: 19h 30' 23,64" (J2000)
DEC: +37° 37' 06,58" (J2000)

This candidate is identified in the GAIA DR2 catalog under the number 2051827137119772032 with the magnitudes (rp) = 17.8, (bp) = 18.1, et (g) = 18.9. The Ra 69 spectrum was generated on September 17, 2017. Hardware configuration and exposure times were the same as for Pre 24 (see section 7.3.1).

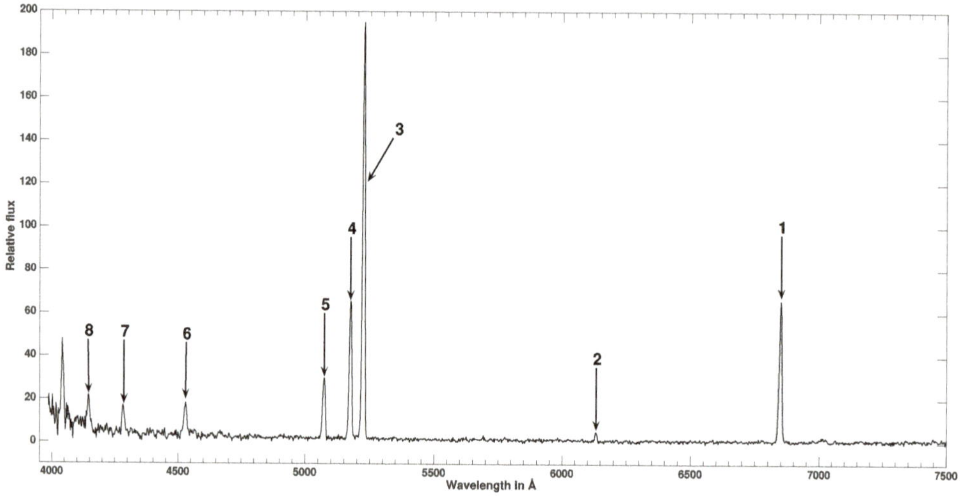

Fig. 7.3.4 — 1D spectrum of Ra 69. Performed with a LISA spectrograph equipped with a 50 μm slit on a 1 m diameter Cassegrain telescope and an ATIK-414EX camera: 4 x 900-second exposures. On the *x*-axis, wavelength in ångströms. On the *y*-axis, intensity relative to the continuum normalized to 1. The numbers indicate the emission lines described in the table of Figure 7.3.5.

As in the case of Pre 24, the spectrum was very similar to that of a planetary nebula, with the predominant presence of the $H\alpha$ and [O III] lines, but here again the spectrum was red-shifted.

This is a galaxy with an active nucleus with even more intense lines than Pre 24 at over 100 times the continuum value for the [O III] 5 007 Å line.

Elements	Line #	λ_o	λ_1	$\Delta\lambda$	Rel. Flux	z
Hα	1	6562,82	6848,71	285,89	98,72	0,04356
HeI 5876	2	5875,62	6132,67	257,05	5,66	0,04375
[OIII] 5007	3	5006,84	5224,92	218,08	128,67	0,04356
[OIII] 4959	4	4958,91	5174,86	215,95	41,91	0,04355
Hβ	5	4861,33	5072,65	211,32	17,77	0,04347
Hγ	6	4340,47	4529,05	188,58	7,16	0,04345
Hδ	7	4101,74	4283,36	181,62	5,76	0,04428
Hε	8	3970,07	4145,41	175,34	6,68	0,04417
					Average z =	Average z

Fig. 7.3.5 — Calculating the redshift of Ra 69. Using the 8 lines identified in Figure 7.3.4. The values of the column "Rel.Flux" correspond to the intensity of the lines measured by modeling a Gaussian curve against the continuum.

We measured an average redshift $z = 0.04372 \pm 0.00051$, a receding velocity of the order of 13 100 km.s^{-1}, which is a distance of 177 Mpc (Approximately 577 million light-years away) using a Hubble constant value of H_0 = 74.03 km.s^{-1}.Mpc^{-1}.

This galaxy is therefore more than 3 times further away than Pre 24.

During our various candidate observation campaigns, we have discovered several galaxies that were previously unclassified which were published with the help of the German astronomer, Michael Koenig.

The publication is available here:

https://arxiv.org/pdf/1906.09612.pdf

7.3.3 Hu 6

Discovered by French amateur astronomer, Laurent Huet. Laurent discovered a dozen candidates on his narrow band filter images. Hu 6 has the following coordinates:

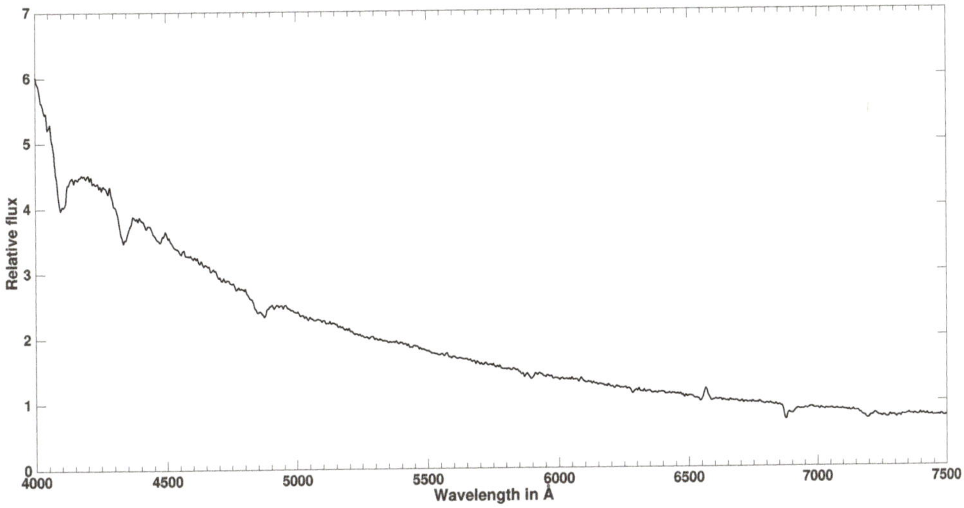

Fig. 7.3.6 — 1D spectrum of Hu 6. Performed with a LISA spectrograph equipped with a 50 µm slit on a 1 m diameter Cassegrain telescope and an ATIK 414EX camera: 2 x 900 seconds of exposure. On the *x*-axis, wavelength in ångströms. On the *y*-axis, the relative intensity with respect to the continuum normalized to 1.

RA: 18h 52' 41,38" (J2000)
DEC: +26° 45' 31,5" (J2000)

Still on the 1-meter telescope at the C2PU on the Calern plateau during the September 2017 mission, we take two 900-second exposures of this fairly bright object.

The spectrum shows the hydrogen lines with the $H\alpha$ line in emission on a continuum that rises into the blue. The other hydrogen lines are not in emission. No significant spectral shift. This time, the target is actually in our galaxy, and corresponds to the spectrum of a fairly hot star. The spectrum was analyzed by Professor Steven Shore of the University of Pisa in Italy who indicated that this is a DA WD dwarf novae which can be found in SIMBAD under the name of V* CY Lyr or 2MASS J18524137+2645314 with a magnitude of $V = 13.20$. It would be interesting to revisit this spectrum to see if the profile has changed

Following this negative spectrum, the baptismal name "Hu 6" was assigned to another Laurent Huet discovery.

7.3.4 DeGaPe 32

Discovered by the APO Team in San Pedro de Atacama, Chile, using their 105 mm Takahashi telescope. The team is composed of Thierry Demange, Richard Galli and Thomas Petit. To date, they have an impressive catalog of 80 new objects discovered from their own images. DeGaPe 32 has the following coordinates:

RA: 05h 17’ 57,10" (J2000)
DEC: +07° 26’ 24,70” (J2000)

The target is located in the Orion constellation, close to the vdB38 nebulae, Sh2-262 and sh2-265. We produced its spectrum on September 20, 2017, again from the C2PU at the Côte d’Azur Observatory, using the same configuration as for the Ra 69 and Pre 24 targets.

The object is bright enough to reveal emission lines. Two exposures of 900 seconds were sufficient.

The spectrum shows the $H\alpha$ line in fairly intense emission at more than 10 times the continuum level. Other Balmer emission lines can also be seen, as well as some helium lines but no [O III] line is visible, so we are probably not in the presence of a planetary nebula. We believe it is a Herbig Ae/Be type star, strongly attenuated by interstellar extinction (the continuum being nearly horizontal).

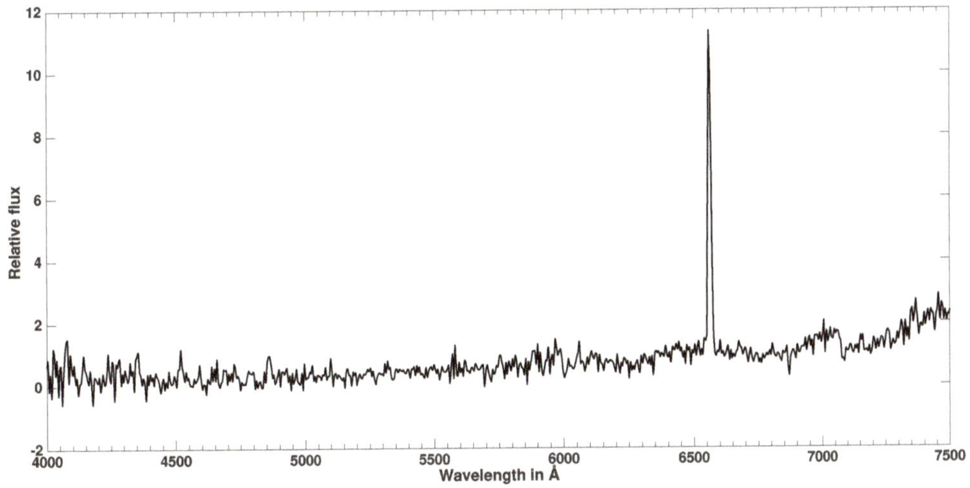

Fig. 7.3.7 — 1D spectrum of DeGaPe 32. Performed with a LISA spectrograph equipped with a 50 μm slit on a 1 m diameter Cassegrain telescope and an ATIK 414EX camera: 2 x 900 second of exposures. On the x-axis, wavelength in ångströms. On the y-axis, intensity relative to the continuum normalized to 1.

We decided to send our spectrum to astrophysicist Coralie Neiner, Be star specialist to get her opinion as well as Professor Steven Shore. Their answers seem to point to an LBV-type B supergiant star, but not a Herbig Ae/Be because, according to Coralie Neiner, the $H\alpha$ line is too thin.

7.3.5 Pa 30, a very strange object...

Discovered by American amateur astronomer Dana Patchick in August 2013, by analyzing images from surveys such as the Wise MIR and the DSH(1). Pa 30 has the following coordinates:

RA: 00h 53' 11,2" (J2000)
DEC: +67° 30' 02,50" (J2000)

This last example remains, from our point of view, the most interesting case, the one that generated the most suspense, but also the one behind a major discovery in astronomy.

It all began during our mission to the Astroqueyras observatory in 2016 where the Pa 30 candidate was one of our targets for our LISA spectrograph mounted on an RC 500 mm Astrosib. We made an initial spectrum by exposing for 1 200 seconds but paid little attention to it as the spectrum showed nothing of note on a single exposure, just a continuum covering the LISA spectral range between 4 000 and 7 500 Å. We noted a rise in the signal close to 4 000 Å, which we attributed to noise, given the magnitude of this target. Then, we moved on to the next target.

In October 2018, Pascal Le Dû decided to take another look at this strange-looking object. In fact, the object is classified as a probable planetary nebula in the HASH database, but it has no spectrum, which is curious... Observation was carried out from his home, at the end of Brittany's tip with a Newton telescope of 200 mm diameter open at F/5. He took 9 exposures of 1 200 seconds, for a total of 3 hours of exposure, obtaining a spectrum showing a spectrum that goes much further into the near ultraviolet than a LISA. This spectrum showed a broad emission peak over 400 Å wide.

Pascal was perplexed, wondering whether the signal observed in the near-UV range was real. He requested Thomas Petit, Christian Buil and myself to take spectra to confirm his observation. The weather was not on my side in October 2018, like Christian's, but Thomas from Prague managed to confirm the signal observed by Pascal. This signal is therefore real, and not due to an artifact at the time of acquisition or processing. Pascal sent the spectrum to astrophysicist Quentin Parker at the University of Hong Kong.

His response was not long in coming and indicated that the signal in blue is surprising, even atypical. The suspect line is identified as that of 5 times ionized oxygen ([O VI] 3811-3834). It is very large and corresponds to a powerful stellar wind of around 15 000 $km.s^{-1}$.

Shortly afterwards, Christian succeeded in observing this object with an UVEX (all-reflecting spectrograph) prototype equipped with a 600 line/mm grating which further detailed the observed signal. On the professional side, a spectrum was observed in 2014, but the spectrograph used at the time, like our observation during our mission to the Saint-Véran observatory in 2016, did not cover a sufficiently broad wavelength range in the near ultraviolet. This broad emission line could not be detected at wavelengths below 4 000 Å by the equipment which was used.

So it was Pascal's spectrum that aroused the curiosity of professionals, who then embarked on new spectral measurements including a spectrograph covering a wider range of near-ultraviolet wavelengths. They obtained much less noisy spectra than Pascal's.

These spectra also revealed the very discreet presence of [S II] lines (6 716 Å and 6 731 Å) in the material shell surrounding the candidate, which show an expansion velocity of the order of 1 100 km.s^{-1}. In Pascal's spectrum, the [S II] line signal is almost drowned out by noise, making precise measurements impossible.

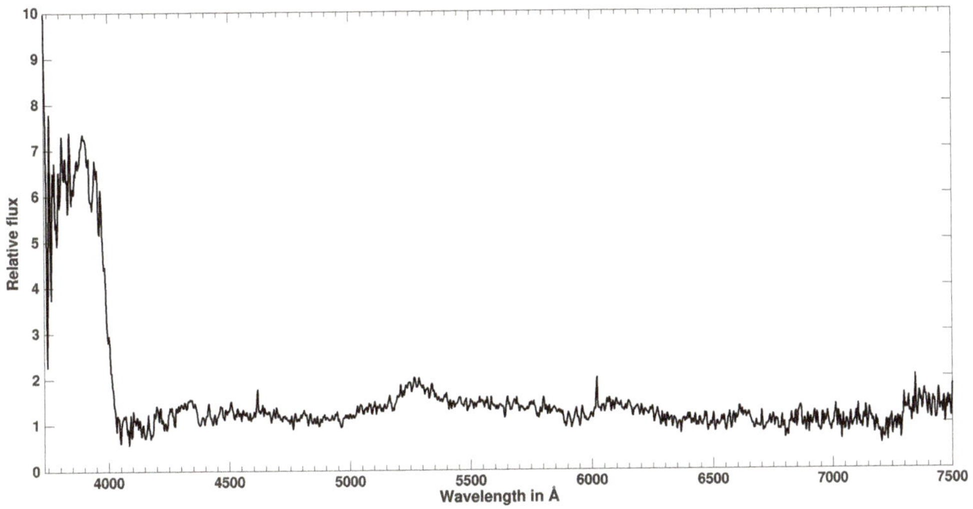

Fig. 7.3.8 — 1D spectrum of Pa 30. Made by Pascal Le Dû with an Alpy 600 and a 200 mm Newton telescope open at F/4. This spectrograph goes further into the near UV than the LISA spectrograph, which has enabled us to better highlight this broad emission between 3 800 and 4 000 Å. On the x-axis, wavelength in ångströms. On the y-axis, intensity relative to the continuum normalized to 1.

Other spectra have also been produced and the studies carried out in this campaign on Pa 30 show that the object is the remnant of a supernova afterglow associated with a very hot central star, the hottest ever observed, at over 200 000 K, making it one of the hottest objects in our galaxy.

The spectrum, produced by Christian Buil using a UVEX (see Figure 7.3.9), shows the full width of the [O VI] line. In this case, we can see how important it is to choose the right spectrograph for the task in hand.

For the Pa 30 candidate, we would have missed a major discovery if no one had followed up by exploring the near-ultraviolet part.

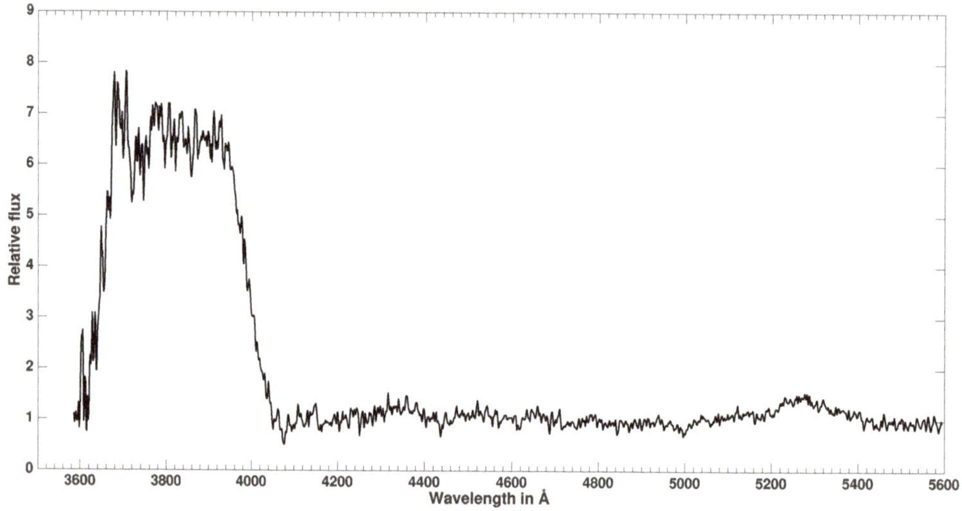

Fig. 7.3.9 — 1D spectrum of Pa 30 with UVEX. Made by Christian Buil with a UVEX equipped with a 600 line/mm grating ($R = 800$ approx.) on a Ritchey-Chrétien telescope of 225 mm in diameter and open to F/8. The spectrum covers a much wider range on the near-UV side than the Alpy 600. On the x-axis, wavelength in ångströms. On the y-axis, intensity relative to the continuum normalized to 1.

What was interpreted as noise on the very first spectrum we obtained with LISA in 2016 (which could not go to wavelengths shorter than 4 000 Å), was revealed to be the beginning of a broad emission line, which the Alpy 600 spectrograph was able to show, and which the UVEX confirmed by going even further into the near ultraviolet. The UVEX has been precisely designed for optimum performance in this range, provided you use a telescope equipped only with mirrors (no lenses), to allow short wavelengths through.

Christian's spectrum shows the [O VI] line in very broad emission, extending over about 500 Å (from 3 600 to nearly 4 100 Å). We can also see that its profile is not symmetrical and contains a lot of information relating to its peak, which is not just noise. The peak structure is undoubtedly the signature of an inhomogeneous stellar wind with lots of dust.

Using the UVEX spectrum, we were able to measure the stellar wind speed generated by Pa 30 and we find a speed similar to that published by Foteini Lykou *et al.* (2023). This publication indicates that it is in fact the supernova that exploded in the year 1181, observed at the time by Chinese and Japanese astronomers. It is a very rare type of supernova of the Iax type resulting from the fusion of 2 white dwarfs.

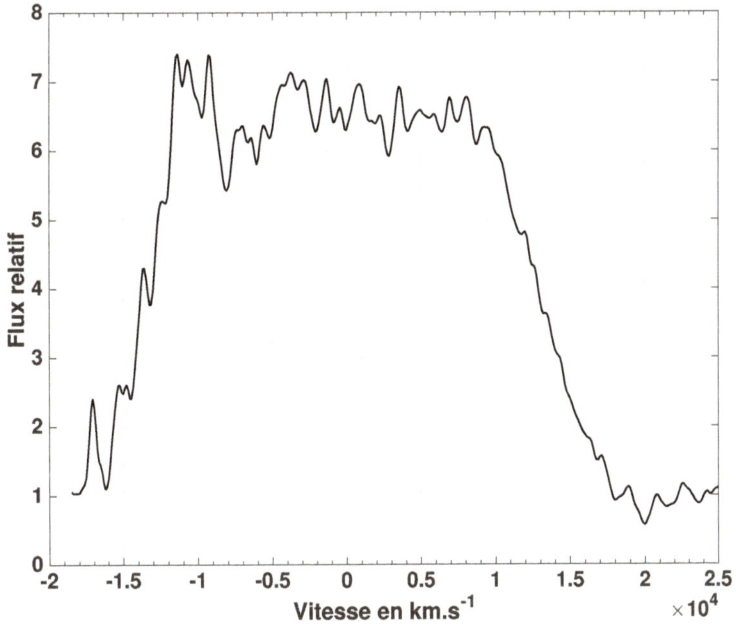

Fig. 7.3.10 — Pa 30 stellar wind velocity. Based on Christian Buil's UVEX spectrum and centered on the broad emission line [O VI] 3811-3834 with a stellar wind of around 15 000 km.s^{-1}. On the *x*-axis, velocity in km.s^{-1}. On the *y*-axis, relative intensity with respect to the continuum normalized to 1.

The adventure linked to this candidate shows on the one hand, that amateurs have played a major role by producing a spectrum that has alerted the professional community, but on the other hand that perseverance pays off in spectroscopy. A universal spectrograph does not exist, you have to look beyond the classic visible wavelength range to occasionally obtain surprising and useful results.

7.4 Bibliography

- The planetary nebulae website: https://planetarynebulae.net/FR/

- Amateur PN discoveries and their spectral confirmation: A significant new addition to the Galactic PN inventory by Pascal Le Dû et al., A&A 666, A152, 2022. DOI: 10.1051/0004-6361/202243393

- Nébuleuses planétaires: joyaux de l'astrophotographie. L'Astronomie, Vol. 128, no. 68, p. 40, Jan 2014

- Le bilan des découvertes par les amateurs en France. L'Astronomie, Vol. 129, part no 1, 42-49, 2015

- De nouvelles nébuleuses planétaires découvertes par des astronomes amateurs. L'Astronomie, Vol. 129, No. 80, pp. 42-49, Feb 2015

- Nouvelles candidates "Nébuleuses planétaires". L'Astronomie, Vol. 131, Iss 102, p. 46-51

- VizieR Online Data Catalog: New planetary nebulae candidates (Le Dû, 2017), Originally published in: 2017LAstr.102...46L; 2017LAstr.131b..46L

- Nébuleuses planétaires: de nouvelles candidates. L'Astronomie, Vol. 132, part no 3, 54-57, 2018

- VizieR Online Data Catalog: New planetary nebulae candidates (Le Dû, 2018), VizieR On-line Data Catalog: J/other/LAstr/114.54. Originally published in: 2018LAstr. 114...54L; 2018LAstr.132c..54L

- Planetary nebulae discovered and confirmed by amateur astronomers by Pascal Le Dû *et al.* (SF2A-2018: Proceedings of the Annual meeting of the French Society of Astronomy and Astrophysics. Eds.: P. Di Matteo, F. Billebaud, F. Herpin, N. Lagarde, J.-B. Marquette, A. Robin, O. Venot, pp. 443-451).

- Discovery of New Faint Northern Galactic Planetary Nebulae by Agnès Acker *et al.* Revista Mexicana de Astronomía y Astrofisica, Vol. 48, pp. 223-233, 2012 (http://www.astroscu.unam.mx/ rmaa/)

- A spectroscopic survey of 51 planetary nebulae by L.H. Aller and C.D. Keyes, The Astrophysical Journal supplement series 65405-428, Novembre 1987

- Discovery of an Exceptional Optical Nebulosity in the Suspected Galactic SN Iax Remnant Pa 30 Linked to the Historical Guest Star of 1181 CE by Robert A. Fesen *et al.* The Astrophysical Journal Letters, 945:L4 (7pp), mars 2023

- A new study on a type Iax stellar remnant and its probable association with SN 1181 by Foteini Lykou *et al.*, The Astrophysical Journal, 2023, DOI:10.3847/1538-4357/acb138.

- The Remnant and Origin of the Historical Supernova 1181 AD by Andreas Ritter *et al.*, The Astrophysical Journal Letters, 918L33 (6pp), Septembre 2021, DOI: 10.3847/2041-8213/ac2253.

- Four newly discovered HII galaxies by Olivier Garde *et al.* (doi.org/10.48550/arXiv. 1906.09612, 2019)

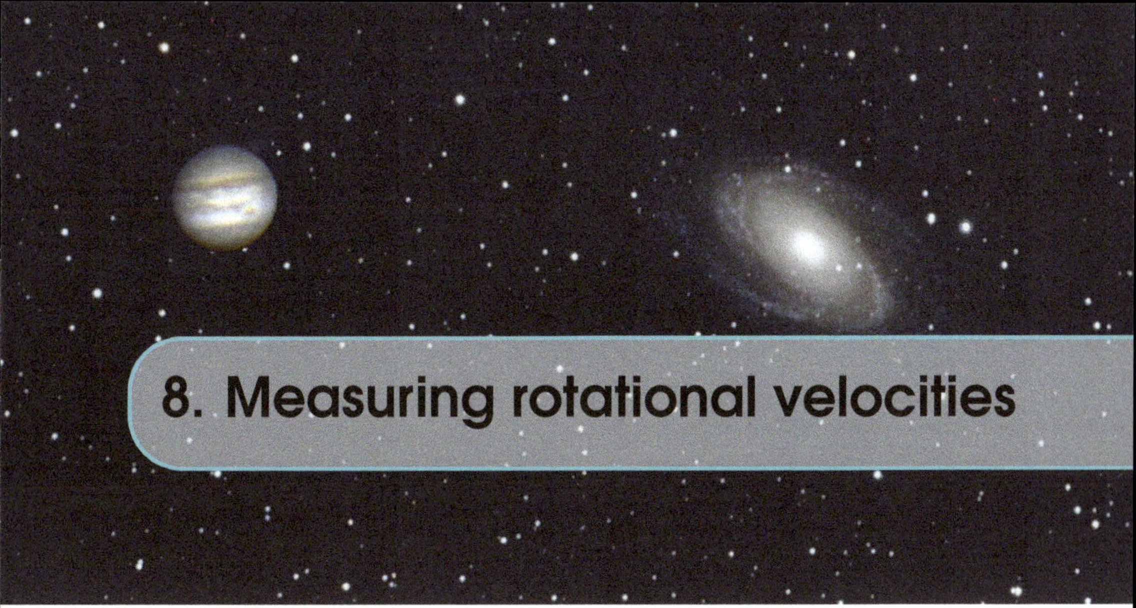

8. Measuring rotational velocities

Thanks to the Doppler-Fizeau effect, spectroscopy allows rotational velocities to be measured by measuring the spectral shift of a line or its width.

Spectral measurement of rotational velocities in astronomy concerns all objects that rotate on their own axis; this can apply to objects as diverse as planets or galaxies and accretion disks around black holes. These velocities allow other parameters to be calculated, such as the mass of the objects and their distance from the central object.

In this chapter, we will discuss a few specific cases as well as the calculation methods appropriate to each situation which can be applied to other objects of the same type.

Warning! In all cases presented, the measured velocities are not the real rotational velocities. This depends on the angle of inclination (i) of the object's axis of rotation along the observer's line of sight. Unless one is fortunate enough to have an axis of rotation that is perfectly perpendicular to the observer's line of sight, which is quite rare!

The velocity measured as a function of the angle of inclination of this axis is:

$$V_{Mes} = V_{Real} \sin(i) \quad \text{and} \quad V_{Real} = \frac{V_{Mes}}{\sin(i)}$$

where:

V_{Mes}: the velocity measured on the spectrum.

V_{Real}: the real rotational velocity.

i: the inclination of the axis of rotation relative to the line of sight.

8.0.1 Influence of the rotation axis inclination

First, let us define what the angle of inclination (i) is.

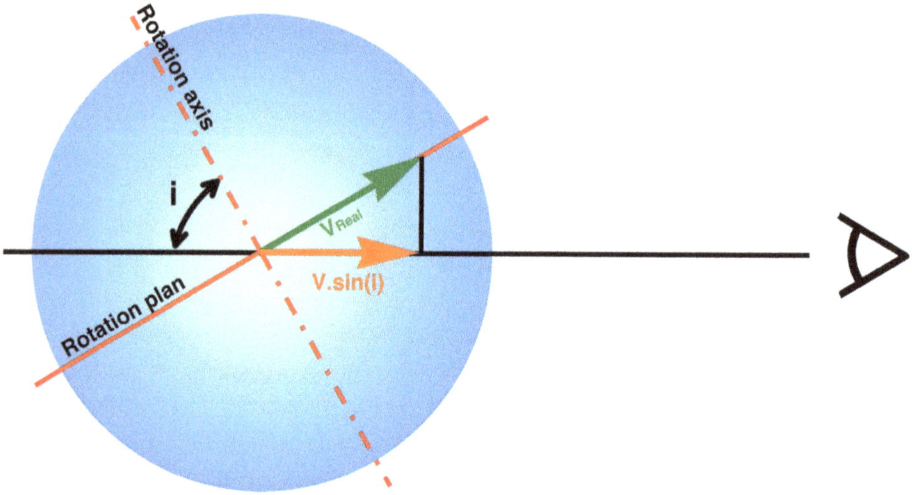

Fig. 8.0.1 — Projection of the rotational velocity of an object. The velocity measured by the observer in the line of sight depends on the inclination (i) from its axis of rotation. The observed velocity is therefore modified by the sine value of the angle of inclination of the axis of rotation.

• If the target has a rotation axis (i) of 90° relative to the observer (perpendicular to the line of sight), then $\sin(90°) = 1$ and the measured velocity is indeed that of the object.

• If the rotation axis of the target is parallel to the observer's line of sight, so an angle of 0°, then $\sin(0°) = 0$ and in this case, it will not be possible to measure velocity using the Doppler-Fizeau effect.

• With intermediate angle values, the measured velocity ($V\sin(i)$) will always be lower than the real rotation velocity.

In practice, it is often difficult to calculate the inclination of the object's axis of rotation. For some galaxies, it is possible to estimate the inclination by modeling an ellipse, but in the case of a point source, it is more complicated, or even impossible. That is why we often talk about $V\sin(i)$, the real velocity cannot be known exactly.

The velocities to be measured vary greatly: of a few tens of km.s^{-1} for the planets in our solar system to a few hundred km.s^{-1} for galaxies, up to several thousand km.s^{-1} for the rotational velocity of a black hole's accretion disk.

This diversity of velocities, but also of magnitudes between the various types of objects, will require the use of different spectrographs depending on the order of the velocities to be measured, but also taking into account the magnitude of the object.

For example, Jupiter's spectrum requires a high-resolution spectrograph, however, it will be difficult to measure the rotational velocity of a galaxy's arms with this same instrument. It is therefore necessary to adapt the instrument appropriately to the magnitude of the target and the velocities to be measured.

In any case, great care must be taken to calibrate the spectra in wavelength.

8.1 The rotation of the planets

All the planets in our solar system rotate at very different velocities. The fastest is Jupiter, with a rotational velocity of around 13 km.s^{-1} and a sidereal day duration of 9 h 55 m ie it rotates on its axis in just under 10 hours.

The slowest is Venus, with a rotational velocity of 0.001 811 km.s^{-1} and a sidereal day lasting more than 243 earth days.

One year on Venus (225 days), the time it takes for the planet to complete one full rotation around the Sun is less than one Venusian day!

The case of Mercury is very similar with a period of revolution around the Sun of almost 88 days and a period of rotation on its own axis in around 59 days.

Mercury and Venus have very slow rotational velocity compared to other planets in the solar system. Amateur instruments cannot measure them.

Name	Diameter in km	Rotational velocity in $km.s^{-1}$	Duration of the sidereal day in hours
Mercury	4 879	0.003	1 407.6
Venus	12 104	-0.0018	-5 832.5
Earth	12 756	0.44	23.9
Mars	6 792	0.24	24.6
Jupiter	142 984	12.66	9.9
Saturn	120 536	10.23	10.7
Uranus	51 118	-4.11	-17.2
Neptune	49 528	2.7	16.1

Fig. 8.1.1 — Rotational velocities and sidereal day lengths of the planets. According to the NASA website. Negative values indicate that the planet rotates in the opposite direction to the others. (Venus and Uranus).

Based on the table in Figure 8.1.1, we can see that the two planets that are easiest to measure are Jupiter and Saturn with respective velocities of 12.66 $km.s^{-1}$ and 10.23 $km.s^{-1}$.

LHIRES III and Star'Ex HR are the most suitable spectrographs for measuring such velocities. It is strongly recommended to use an instrument equipped with a slit rather than a fiber optic. The measurement will be taken by positioning the slit on the equatorial zone of the planet.

8.2 Practical case study with Jupiter

Measuring Jupiter's rotation velocity is a very easy experiment to perform when using the right high-resolution spectrograph and equipped with a thin slit. The order of rotational velocity to be measured and the brightness of this very luminous planet make it an ideal target for beginners.

If you are using an LHIRES III or Star'Ex HR spectrograph, a more narrow slit will result in a higher spectral resolution. But, an more narrow slit means you will need to expose for a longer time to obtain the same signal to noise ratio on the spectrum. This is not necessarily a problem since the planet is bright!

The spectrograph will be adjusted by centering the $H\alpha$ wavelength on the sensor. This should be done during the day using sunlight.

The choice of the $H\alpha$ line is justified here in order to verify the wavelength calibration thanks to the presence of atmospheric lines on either side of the line.

Then, at night, the slit must be aligned with Jupiter's equator by rotating the spectrograph at the exit of the optical tube. It is imperative that the slit be located on the planet's equatorial band.

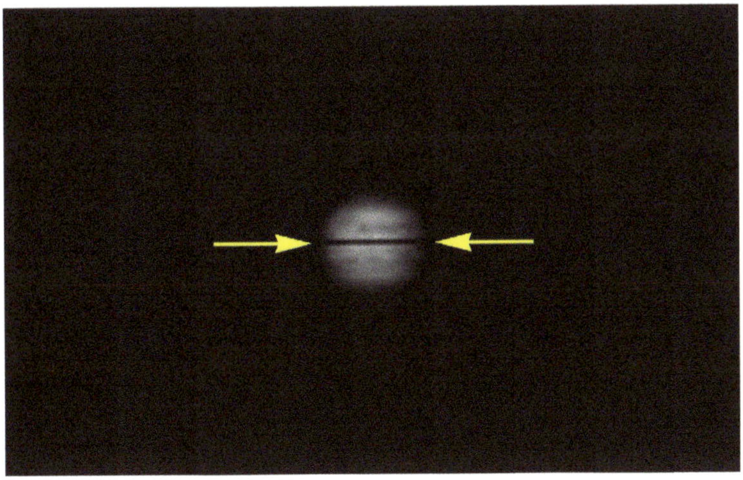

Fig. 8.2.1 — Centring Jupiter on the slit of the spectrograph. Image taken with a 400 mm diameter Ritchey-Chrétien telescope at F/8 (3 200 mm focal length) and an ASI 174 mini camera installed on the autoguiding output of the spectrograph. The two yellow arrows indicate the position of the slit centered on the equatorial band of the planetary disk.

On planets smaller than Jupiter or Saturn, a high focal length is required to obtain a spectrum of a disk with a significant diameter. Ritchey-Chrétien or Schmidt-Cassegrain telescopes are preferable in these cases.

A series of acquisitions is launched by determining a unit exposure time to obtain a signal of several tens of thousands of ADUs in order to optimize the signal-to-noise ratio of the spectrum while maintaining a safety margin to avoid saturation. At the end of the acquisitions, the spectrum of a calibration lamp (argon/neon, for example) will be performed.

The height of the spectrum produced corresponds to the diameter of Jupiter. We can see that some of the lines are slanted. The broad line in the center indicated by two red arrows in Figure 8.2.2 is the $H\alpha$ line reflected by the planet. The spectrum is of light from the sun reflected back to the observer from Jupiter.

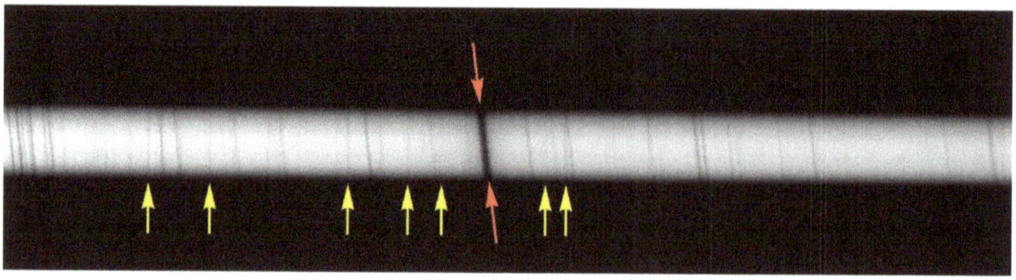

Fig. 8.2.2 — Raw 2D spectrum of Jupiter. Produced using a LHIRES III spectrograph mounted on a 400 mm diameter Ritchey-Chrétien telescope (3 200 mm focal length) and an ASI 533 MM pro CMOS camera: 4 exposures of 180 seconds.

Other lines, however, are perpendicular to the spectrum continuum (marked with yellow arrows). These are the lines of the Earth's atmosphere. They will be very useful for calibrating the wavelength spectrum and verifying the accuracy of the calibration.

The treatment will consist of two measurements whose binning zone will be located on both sides of the planet (at the top and bottom of the track).

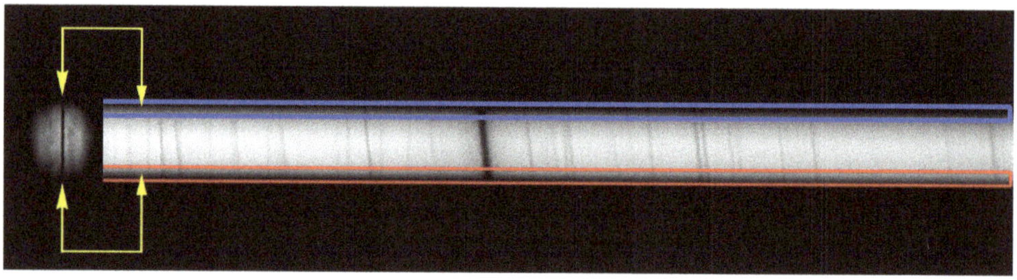

Fig. 8.2.3 — Position of the 2 binning zones. The blue area measures the part of the spectrum shifted towards blue (side moving toward the observer), The red zone is shifted toward red (away from the observer).

The binning area of each edge should be as narrow as possible and well positioned at the edge of the planetary disk. Adjusting the spectrum display thresholds can help to clearly define these two areas.

After processing the two spectra separately, a wavelength calibration check can be performed simply by superimposing them, which will allow us to verify whether the atmospheric lines are properly aligned. If this is not the case, one of the two spectra will need to be corrected by translation. ISIS does it very well!

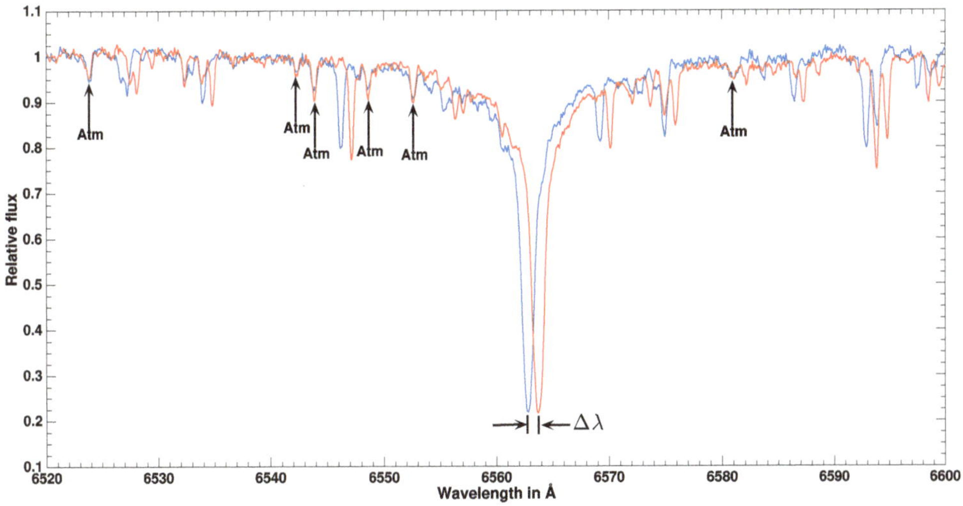

Fig. 8.2.4 — Jupiter spectra. The blue spectrum was produced at the edge of the planet as it approached the observer, the red on the edge furthest from the observer. We can see that the atmospheric lines are superimposed. On the *x*-axis, the wavelength in ångströms. On the *y*-axis, the relative intensity compared to the continuum normalized to 1.

Figure 8.2.4 highlights the difference between the two edges of the planet's disk noted $\Delta\lambda$ and corresponds to:

$$\Delta\lambda = \lambda_{red} - \lambda_{blue}$$

Where:

λ_{red}: the value in ångströms of the red-shifted $H\alpha$ line.
λ_{blue}: the value in ångströms of the blue-shifted $H\alpha$ line.
$\Delta\lambda$: the value of the shift of the two lines in ångströms.

The two values of the $H\alpha$ line position are measured according to the explanations in section 3.8. The value of $\Delta\lambda$ measured in this example on Jupiter is 1.08 Å. To determine the planet's rotational velocities, the following equation is used:

$$V\sin(i) = \left(\frac{c}{4}\right)\left(\frac{\Delta\lambda}{\lambda_0}\right)$$

With:

$V\sin(i)$: the rotational velocity according to inclination (i) of the rotation planet's axis relative to the observer. The inclination i, which depends on the observation period, can be found on the website of the Institute of Celestial Mechanics and Ephemeris Calculation (IMCCE in Paris-France):

https://www.imcce.fr/

c: the speed of light (299 792.458 km.s^{-1}).
$\Delta\lambda$: the value of the shift in ångströms.
λ_0: the value of the reference wavelength at rest in ångströms. If the $H\alpha$ line is used, this value is 6 562.82 Å.

The division coefficient of 4 may seem surprising, but it is important to remember that the Doppler-Fizeau shift effect occurs four times in this case:

• Twice, because we measure the velocity on both sides of the planet's disk

• Twice more, because the light is not emitted by the planet but reflected and the Doppler-Fizeau effect therefore acts twice (once when the light reaches the planet, and again when the light is reflected back to the observer).

If we apply the numerical calculation using the measurement determined previously: $\Delta\lambda$ = 1.08 Å, we obtain a rotational velocity $V\sin(i)$ = 12.33 km.s^{-1}, a value fairly close to the table in Figure 8.1.1. This measurement gives an order of magnitude for Jupiter's rotational velocities.

The calculation example in this chapter clearly shows that the result is consistent, but it is not an absolute value.

Measurement errors can quickly arise. Firstly, in the selection of the binning area, where it can be difficult to accurately determine the real edge of the planet. Furthermore, turbulence during acquisitions can cause variations in the diameter of Jupiter's disk.

Jupiter's rotation axis is tilted relative to the observer on Earth except during the Jovian equinoxes, when the measurement of its rotational velocities would be more accurate because the axis of rotation appears perpendicular to the axis of the terrestrial observer. In reality, the inclination of Jupiter's axis of rotation on the ecliptic is 3.1 degrees, which means that the angle has little effect on the velocity measurement (sin(90-3.1°) = 0.998).

The Jupiter phase can also cause errors if this type of measurement is not taken during a period of opposition. Poor seeing during acquisitions can also influence the results produced.

Finally, Jupiter is a gas planet: the boundary between its atmosphere and space is not clearly defined.

8.3 The rotation of Saturn and its rings

The same measurements can be taken on Saturn, which remains a relatively easy target both because of its high rotational velocities around 10 km.s^{-1}, and thanks to its large apparent diameter. It will even be possible to measure the rotational velocities of the rings, but you will need a long focal length and very good seeing to separate them clearly from the planetary disk.

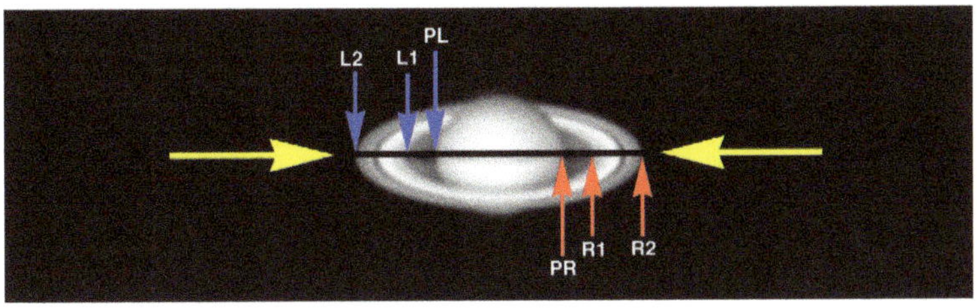

Fig. 8.3.1 — Centring the slit on Saturn. The position of the slit is indicated by the two yellow arrows. The blue arrows show the rotating part moving toward the observer, the red arrows indicate the parts that are moving away. Image taken with the autoguiding camera on an LHIRES III spectrograph installed on a Ritchey-Chrétien telescope with a diameter of 500 mm and a focal length of 4 meters at the Saint Véran observatory.

The spectrum can be obtained using a high-resolution spectrograph such as the Star'Ex HR or the LHIRES III equipped with a 2 400 lines/mm grating and a 23 μm slit or even 10 μm, which increases spectral resolution. Saturn's high brightness allows for short exposure times.

The spectrum processing is similar to that performed for Jupiter. The measurements will be taken on the $H\alpha$ line, at various points on the planet and its rings, as shown in the Figure 8.3.1.

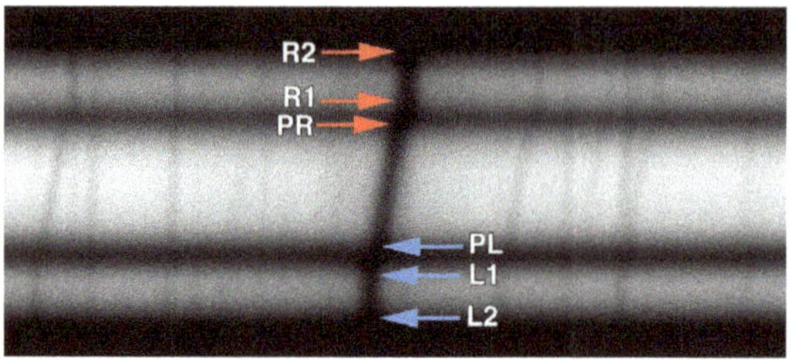

Fig. 8.3.2 — Raw 2D spectrum of Saturn. Note the difference in the inclination of the $H\alpha$ line between the rings and the planet. Spectrum produced during the AstroMaths/ENS Lyon camp by Daniel Vérilhac with the RC 600 at F/D = 8 from the Mars Observatory in Ardèche - France. LHIRES III spectrograph, Atik 460EX camera. 300 second exposure.

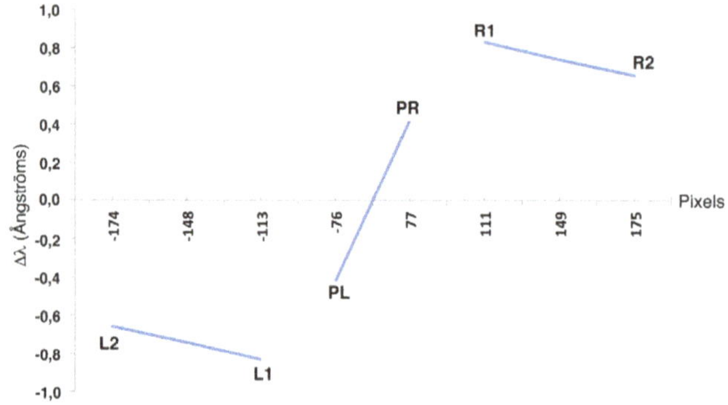

Fig. 8.3.3 — Keplerian rotation of the rings. Deviations in ångströms ($\Delta\lambda$) relative to the rest $H\alpha$ line for the points indicated in Figure 8.3.1. On the x-axis, the distance in pixels from the center of the planet.

For each spectrum, a wavelength calibration check will be performed using the atmospheric lines present around $H\alpha$, which can also be used to calibrate with greater precision because they are already present in the acquisition.

We can see that the rings do not rotate at the same velocity as the planet and that the ring closest to the planet rotates faster than the outer ring (Just like the International Space Station in low Earth orbit rotates faster than a geostationary satellite located at an altitude of 36 000 km).

This is actually a Keplerian rotation, which demonstrates that the rings are not formed from a "rigid block" but from a collection of small independent bodies all orbiting together around the planet.

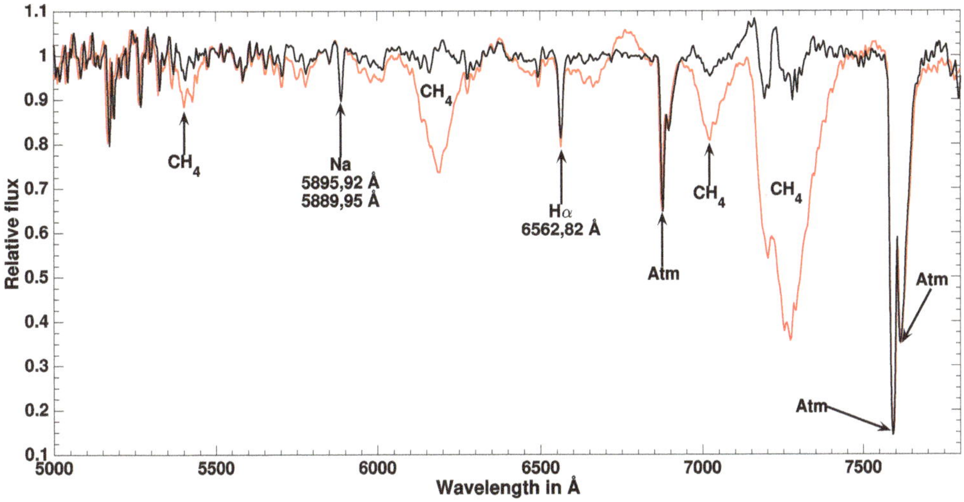

Fig. 8.3.4 — The chemical nature of Saturn. Difference in spectral profile between the planetary disk (in red) with the presence of molecular bands CH_4 and the rings (in black). Spectrum produced during the AstroMaths/ENS Lyon camp by Daniel Vérilhac with the RC 600 at F/D = 8 from the Mars Observatory in Ardèche - France. LISA Spectrograph with 35 µm slit, Atik 460EX camera. 5 x 5 seconds. On the x-axis, the wavelength in ångströms. On the y-axis, the relative intensity compared to the continuum normalized to 1.

A low-resolution spectrum shows that they are not made of the same material as the planet, which exhibits absorption lines of methane and ammonia. The spectrum of the rings is practically the same as that of sunlight, so we can assume that they are partly made of ice!

These results are, of course, fairly approximate, as explained for the measurements of Jupiter and do not take into account the inclination of the Saturn axis of rotation relative to the observer on Earth. It should be possible to calculate the inclination (i) of Saturn and deduce its real rotational velocity. This can be done by modeling the boundaries with an ellipse.

Measurements on other gas planets, Neptune and Uranus, may be considered, but these will need to be very precise because the rotational velocities are three to four times less significant than for Jupiter and Saturn. The axis of rotation of Uranus is highly inclined (97.77 degrees), making measurement very difficult.

For terrestrial planets with a "visible surface" (Mars and Mercury), measuring rotational velocities using the Doppler-Fizeau effect is not feasible in amateur spectroscopy due to their low rotational velocities (for example 240 $m.s^{-1}$ for Mars). For Venus, since the surface is not visible, we measure the velocity of the upper layers of the atmosphere, not that of the planet itself!

8.4 Full width at half maximum of a line (FWHM)

A good indicator of an object's rotation is the measurement of the width of a spectral line at half height, which is denoted as FWHM. The unit of width can be expressed in ångströms or in $km.s^{-1}$.

It is obtained by measuring the width of the line at half its maximum intensity (height of the peak of the line relative to the spectrum continuum).

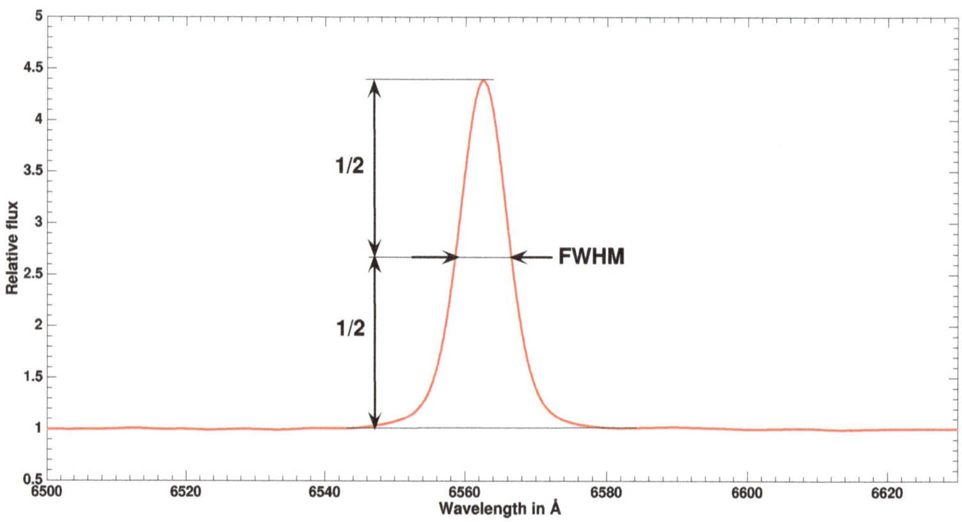

Fig. 8.4.1 — Measurement of width at mid-height. The $H\alpha$ line of a Be star. On the x-axis, the wavelength in ångströms. On the y-axis, the relative intensity compared to the continuum normalized to 1.

When the rotational velocities increases, the FWHM of the line also increases. Indeed, when the difference in peripheral velocities between the side approaching and the side moving away from the observer becomes increasingly greater, the Doppler-Fizeau effect increases the initial wavelength toward the red for the receding part and reduces it for the approaching part (blue). This results in a widening of the resulting line: the higher the rotational velocities, the greater the width.

But be careful! The width of a line is not solely due to the rotational velocity; many other physical parameters also come into play, such as:

- The density of the object and the interactions between atoms.
- Its temperature, which also increases the velocities of the atoms.
- Its pressure, the shocks are more numerous.
- The degree to which it is excited by an external source.
- The turbulence of the matter constituting the observed object (atmosphere, arms).

Depending on the magnitude of the velocities to be measured, it will therefore be necessary to take into account the physical characteristics of the object being studied. We will not cover all possible measurements depending on the type of object in this book, but let's just cite two very different examples in terms of the velocities to be measured:

- The rotational velocity of the disk surrounding a Be star type.
- The rotational velocity of a disk of matter near a black hole at the center of a galaxy.

If you want to measure the rotational velocity of another type of object, we can refer to various publications by searching through scientific literature. It should also be noted that there are no general rules on how to measure rotational velocities, Several methods exist, and not all of them involve measuring the width of a line.

First, we will see how to measure the full width at half maximum of a spectral line. Several software programs, such as ISIS and Visual Spec, enable this measurement. Here is how to proceed in the Visual Spec software:

1) In the menu "**File**", select "**Open profile**". Click to the left of the profile to be measured and while holding down the left mouse button, move the mouse cursor to the right of the profile to frame the line to be measured. Then make a "**right click**" with the mouse, which displays a pop-up menu where you must select "**Computations**".

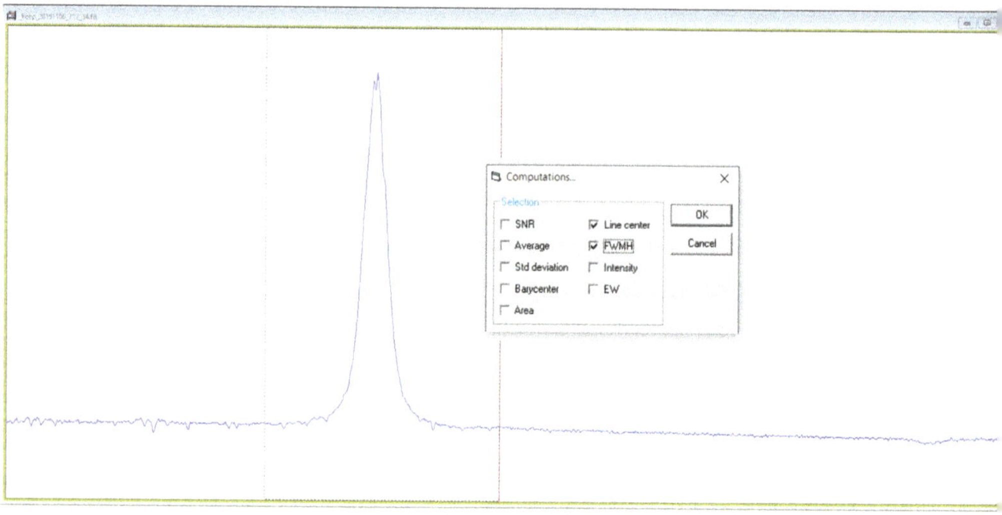

Fig. 8.4.2 — Measurement of FWHM. With Visual Spec software.

2) In this local menu (Figure 8.4.2), you have the option to select several possible measures, check "**Line center**" and "**FWHM**", then press OK.

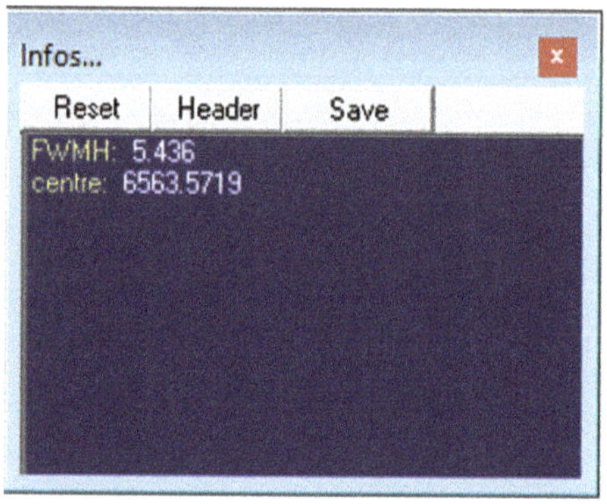

Fig. 8.4.3 — Result of the FWHM measurement. With Visual Spec software.

3) In return, you will receive the two measurements requested (Figure 8.4.3), namely the full width alf maximum (FWHM) and the position of the center of the line (in ångströms).

The measured FWHM value must be corrected by the instrumental FWHM value of the spectrograph used. This depends on the resolution R of the spectrograph, that is to say the FWHM of a calibration line or, better yet, an atmospheric line known to be very narrow.

The relationship that allows the measured FWHM to be corrected is as follows:

$$FWHM_{Cor} = \sqrt{FWHM_{Mes}^2 - FWHM_{Inst}^2}$$

With:

$FWHM_{Cor}$: the corrected value of the measured FWHM.

$FWHM_{Mes}$: the value of the FWHM measured on a line of λ wavelength.

$FWHM_{Inst}$: the value of the instrumental FWHM which depends on the resolution of the spectrograph.

There are two possible scenarios: either we know the resolution R for a given wavelength, or we do not know it...

Case 1: Resolution *R* is known

Apply the equation (inverse of the resolution calculation):

$$FWHM_{Inst} = \frac{\lambda}{R}$$

With:

λ: the wavelength of the line for which you want to know the FWHM.

R: the resolution of the spectrograph.

Case 2: The resolution is unknown

The FWHM of an atmospheric line can be measured around the $H\alpha$ line, where water vapor lines are very numerous. It is also possible to use the auroral line [O I] at 5 577.35 Å or other atmospheric lines.

The FWHM of a calibration lamp line can also be measured (but this will be less accurate than with a line from the Earth's atmosphere).

8.5 Measurement of rotational velocities based on its FWHM

The calculation of the rotational velocities for a line is performed by applying the Doppler-Fizeau equation:

$$V\sin(i) = c\left(\frac{FWHM_{Cor}}{\lambda_0}\right)$$

With:

$V\sin(i)$: the value of the rotational velocity in km.s^{-1} measured according to the angle of view (i).

c: the speed of light in km.s^{-1} (299 792.458 km.s^{-1}).

$FWHM_{Cor}$: the value in ångstroms of the full width at half maximun of the corrected instrumental FWHM.

λ_0: the wavelength value of the center of the line whose rotational velocity is to be determined.

Here is a numerical example with a FWHM measured on a spectrum obtained with a spectrograph with a resolution R = 10 000 on the $H\alpha$ line: $FWHM_{Mes}$ = 7 Å.

The value of the FWHM instrumental ($FWHM_{Inst}$) is:

$$FWHM_{Inst} = \frac{6\,562.82}{10\,000} = 0.656 \text{ Å}$$

The FWHM can also be measured directly on the spectrum, either an atmospheric line or a line known to be fine from a calibration lamp. The corrected FWHM value is then calculated ($FWHM_{Cor}$):

$$FWHM_{Cor} = \sqrt{7^2 - 0.656^2} = 6.969 \text{ Å}$$

Finally, we calculate $V\sin(i)$:

$$V\sin(i) = 299\,792.458\left(\frac{6.969}{6\,562.82}\right) = 318.347 \text{ km.s}^{-1}$$

Reminder: the measured value $V\sin(i)$ must be divided by two to obtain the rotational velocity, because the speed is taken into account twice: once for the part of the spectrum shifted towards red that moves away from the observer, a second time for the part of the spectrum shifted towards blue that is approaching. We do not divide by 2 again because this is not reflected light as in the case of planets, but light emitted by the object.

This value is often referred to in some publications by the term HWHM (*half width at half maximum*). It is also not the actual rotational velocities, but rather the speed measured at an angle (i) relative to the observer's line of sight.

8.6 The rotation of a Be star's disk

In Chapter 5, we discussed the physical properties of Be-type stars which have one or more emission lines indicating the presence of a disk rotating around the star at some point in their lives. These emission lines will enable us to calculate the rotational velocity of the disk.

There are several ways to calculate this value.

The numerous publications on this subject show that there is no universal way to perform this calculation. (The bibliography at the end of this chapter will show you some examples). Furthermore, as the physics surrounding these stars is still poorly understood, research continues...

Here we will describe a simple, empirical method that does not claim to be a rigorous, precise or absolute way of measuring this velocity, Rather, it indicates an order of scale whose accuracy will depend on both the inclination (i) of the disk relative to the observer, but also other parameters such as the star's magnetism, turbulence, density, or temperature.

Spectral resolution will also contribute to the accuracy of the results. The choice of lines to use for measuring the rotational velocities is also important, we will see why a little later on.

The method described is based on the assumption that the disk is Keplerian, that is, it obeys Kepler's laws for an object orbiting another as planets do around the Sun, their satellites, or Saturn's rings.

The material in the disk orbiting close to the star will have a higher rotational velocity than the outer material. But in reality, the disc is "viscous". Therefore, the strict application of Kepler's laws is not entirely accurate if one is seeking high measurement precision, but they remain valid as a first approximation.

Here is an example with the Be star 28 Tau (also known as Pleione) located in the Pleiades star cluster (M45). It has a magnitude $V = 5.09$ which makes it easily accessible with a high or medium resolution spectrograph. The use of a low-resolution spectrograph may be considered, but the measurements will be much less accurate, with a margin of error of around $\pm$ 120 km.s^{-1}.

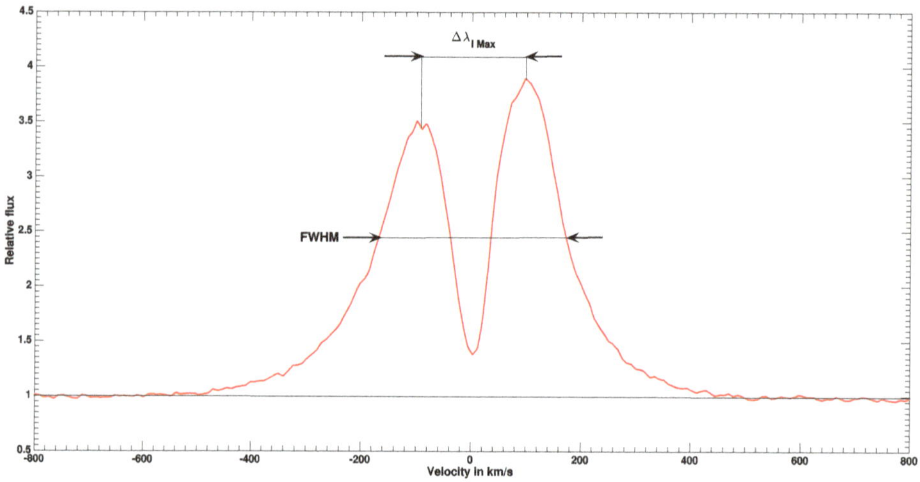

Fig. 8.6.1 — Spectrum of 28 Tau centered on the $H\alpha$ line. Produced with an eShel spectrograph R = 11 000 Mounted on an RC400 telescope, 8 x 300-second exposures. The spectrum was corrected for barycentric velocity and atmospheric lines were removed. On the x-axis, velocities in km.s^{-1}. On the y-axis, the relative intensity compared to the continuum normalized to 1.

The spectrum of 28 Tau shows two emission peaks, The disc is therefore seen almost edge-on along the line of sight (See explanations of emission spectral profiles in Chapter 5 on Be stars). The example was not chosen at random, as the presence of the two peaks will be very useful for certain measurements.

Note that Figure 8.6.1 shows a spectral profile calibrated in velocity rather than wavelength. The central value (0 km.s^{-1}) corresponds to the value "at rest" of the $H\alpha$ line (i.e., 6 562.82 Å). The velocity can therefore be read directly based on the distance from this central value.

On this profile, we will be able to measure:

• The full width at half maximun of the profile (FWHM).

• The distance between the two emission peaks ($\Delta\lambda_{Imax}$) ($Imax$: maximum intensity).

In the rotating disk of a Be star, we do not measure a single velocity but several, depending on the diameter of the disk. The material in the disk close to the star rotates faster than the material located on the periphery of the same disk as we have seen on Saturn's rings (ignoring the fact that the disc is viscous and that in reality the calculation is much more complex to perform).

The measurement of the full width at half maximum of the two lines (FWHM) indicates the average rotation velocity of the entire disk. However, measuring the distance between the two emission peaks indicates the rotation velocity of the portion of the disk where the density of matter is greatest.

Taking the numerical values measured on the spectral profile of 28 Tau, for the $H\alpha$ line, we obtain:

$FWHM_{Cor}$ = 7.42 Å, either $V\sin(i)$ = 338.95 km.s^{-1}, so an average rotational velocity of 169.47 km.s^{-1}.

$\Delta\lambda_{Imax} = \lambda_{PicR} - \lambda_{PicV}$ = 6 564.95 - 6 560.75 Å = 4.2 Å, either $V\sin(i)$ = 191.86 km.s^{-1}, This gives a rotational velocity of 95.93 km.s^{-1} to the densest position on the disk.

We could also measure the rotational velocity of the part closest to the star (the fastest), taking the total width of the line, but in the case of a spectral profile centered on the $H\alpha$ line, The calculation will be incorrect because the Balmer lines, and particularly $H\alpha$, have very broad blue and red wings (so-called optically thick lines).

In this case, it is advisable to use optically thin spectral lines such as those of helium (He I 5 876 Å) or iron (Fe II 5 317 Å). But they are often faint and can easily be confused with the star's continuum, or have a low signal-to-noise ratio, which makes the measurement more random and imprecise.

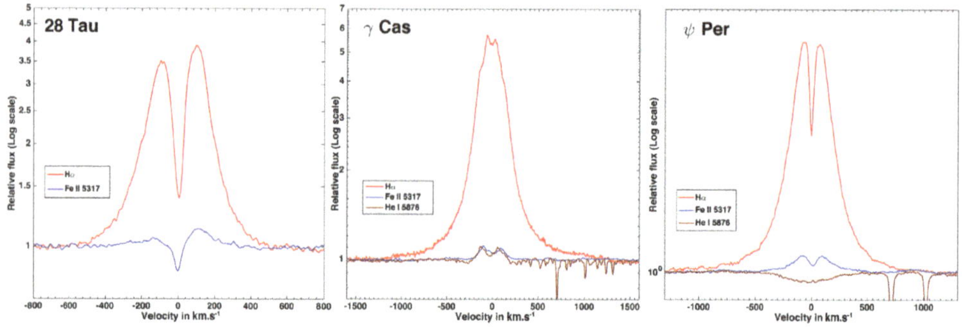

Fig. 8.6.2 — Comparison of spectral lines of Be stars. The red $H\alpha$ line shows very wide wings unlike the Fe II (in blue) and He I (in brown) lines. Spectra produced with an eShel spectrograph(R = 11 000) for 28 Tau and ψ Per, Whoopshel (R = 30 000) for γ Cas. On the x-axis, the velocity in km.s^{-1}. On the y-axis, the relative intensity compared to the continuum normalized to 1 (logarithmic scale).

Here are the wavelength measurements taken on the spectrum of 28 Tau using the $H\alpha$ and Fe II 5 317 Å lines:

Valeur λ	FWHM	$\Delta\lambda_{Imax}$	$\Delta\lambda_{Total}$
$H\alpha$	7.42 Å	4.2 Å	21.65 Å
Fe II 5317	6 Å	3.6 Å	15.75 Å

Fig. 8.6.3 — Table of measurements on 28 Tau. With full width at half maximum (FWHM) values, the difference between the two emission peaks ($\Delta\lambda_{Imax}$) and the total width of the lines ($\Delta\lambda_{Total}$).

The full width at half maximum (FWHM) value between the two lines is similar if converted to velocity:

- With a FWHM on the $H\alpha$ line of 7.42 Å, we obtain a velocity of 338.95. km.s^{-1}.

- With a FWHM measured on the Fe II 5 317 Å line of 6 Å, we find a velocity of 338.30. km.s^{-1}.

This difference of 0.65 km.s^{-1} is negligible in the context of our measurement because it is close to the instrumental error of the spectrograph. However, the difference in the total width of the line on the two profiles is far from negligible:

• With the total width measured on the $H\alpha$ line of 21.65 Å, we obtain a velocity of 989 km.s^{-1}.

• With the total width measured on the Fe II 5 317 line of 15.75 Å, we obtain a velocity of 888 km.s^{-1}.

That is a difference of more than 100 km.s^{-1} ($\approx$ 11% difference).

Consequence: if we want to measure the rotational velocity of the part of the disk closest to the star, the $H\alpha$ line should not be used.

There is also a disparity in velocities if we use the difference between the two peaks in emission with a rotational velocity difference of around 10 km.s^{-1} between the $H\alpha$ and Fe II 5 317 lines, which is still acceptable in the context of this amateur study where we seek to obtain an order of scale for the velocity without calculating it absolutely.

Based on these results, we can calculate/estimate the distance of the disk from its star. W. Hummel and M. Vrancken published in 1995 in the journal *Astronomy & Astrophysics* (Issue 302, pages 751-764) a method for calculating the distance between the star and the disk using the difference between the two peaks in the emission lines of Fe II or He I:

$$\frac{R_d}{R_*} = \left(\frac{2V\sin(i)}{\Delta V_p}\right)^2$$

With:

$V\sin(i)$: the rotational velocity calculated from the full width at half maximum (FWHM) of the emission profile, expressed in km.s^{-1}.

ΔV_p: the wavelength difference between the two peaks, converted into velocity (km.s^{-1}).

$\frac{R_d}{R_*}$: the ratio between the radius of the densest disk (R_d) and the radius of the star (R_*). If we know the radius of the star based on its spectral class, we can determine the radius of the disk.

If, for example, you get a ratio of $\frac{R_d}{R_*} = 3$, The radius of the most intense region of the disk is three times greater than the radius of the star. Similarly, you can measure the ratio between the inner radius (closest to the star) of the disk and the radius of the star, the equation becomes:

$$\frac{R_i}{R_*} = \left(\frac{2V\sin(i)}{\Delta V_{Tot}}\right)^2$$

With:

ΔV_{Tot}: the total width of the line, converted into velocity in km.s^{-1}.

$\frac{R_i}{R_*}$: the ratio between the radius of the disk closest to the star (R_i) and the radius of the star (R_*).

Other measurements can be performed on profiles with only one emission peak. We will not describe the method, which is perfectly explained in the publication by W.Hummel and M. Vrancken (1995): "Non-axisymmetric Be star circumstellar disks".

8.7 The rotation of matter near a black hole in a galaxy

Active galactic nuclei have the particularity of having a spectrum in which certain broad lines are in emission and are the signature of a fast rotating disk near the central black hole in the galactic core. In the spectral profiles of Type 1 Seyfert galaxies, which we discussed in Chapter 3 of this book, there are two types of emission lines:

• The narrow lines such as [O III] lines, which are located in fairly diluted and low-density areas of the galactic core: this is the NLR area (*Narrow Line Region*).

• The broad lines, mainly hydrogen lines, which are very close to the center of the galaxy: this is the BLR area (*Broad Lines Region*).

The NLR region is further from the center than the BLR region. In the case of Type 1 Seyfert galaxies, the observer's line of sight allows them to see the BLR region, so the galactic center and its accretion disk is at an angle between 30 and 60 degrees of inclination.

Type 2 Seyfert active galactic nuclei only exhibit narrow lines because the viewing angle only allows you to see the lines in the NLR area, which masks the BLR region.

It is these broad lines in the BLR zone that will interest us in calculating the rotational velocity of the disk of matter orbiting around the central black hole of these galaxies. It is therefore essential to center the galaxy's core precisely on the spectrograph slit. (Setting the display threshold on the autoguiding camera is very useful for positioning it correctly into the slit).

When processing the spectrum, it will also be necessary to clearly define the binning area on the core and don't hesitate to take a low height by selecting only the most intense area of the core.

To obtain the rotational velocity of the BLR zone disc, the width is measured at half the height of a broad emission line to obtain the value $V\sin(i)$ of the velocity in km.s^{-1}. The spectrum is first corrected for the redshift value of the galaxy to obtain a spectral profile with lines "at rest".

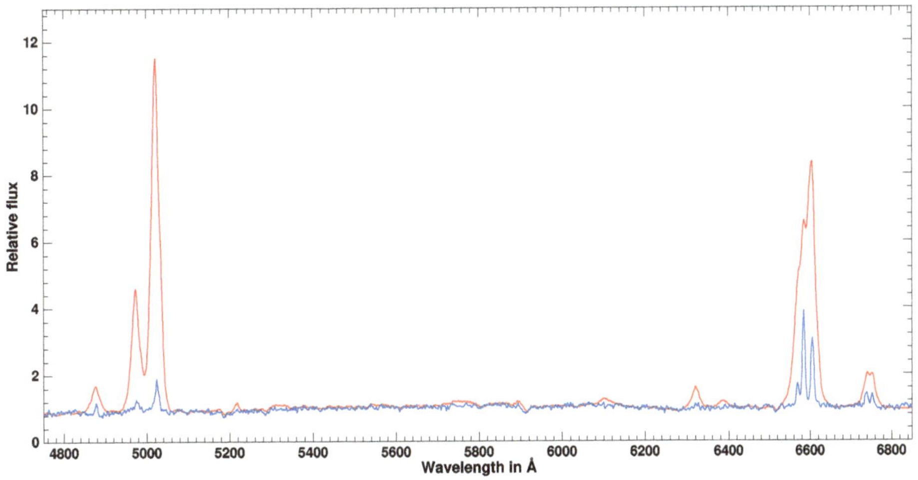

Fig. 8.7.1 — Comparison of spectral zones in a Seyfert 1 galaxy. The blue curve was taken in an area outside the center of the galaxy. It shows finer lines than the red curve centered on the galactic core. Spectra produced using a LISA spectrograph ($R = 780$) on a C14 at F/7 and an ATIK 414Ex CCD camera. On the x-axis, the wavelength in ångströms. On the y-axis, the relative intensity compared to the continuum normalized to 1.

For better accuracy, we can also correct the spectrum for heliocentric velocity (Section 3.7) and galactocentric velocity (Section 3.12) and take into account the instrumental FWHM, which can be corrected using the equation given in section 8.4.

Figure 8.7.1 shows that the width of the lines in the galactic nucleus is very significant. In the case of the $H\alpha$ line, it also encompasses the two [N II] lines located on either side, so we only see the result of these three lines (*blend* = mixing of lines). If we measure the FWHM on the mixture of these lines, the result will not be accurate. The $H\alpha$ line should therefore be avoided when measuring the rotational velocity of an accretion disk. The $H\beta$ line is much more isolated from the other emission lines. Many publications use it for measurements. The disadvantage is that it is less intense than the $H\alpha$ line. However, this line is located at a wavelength that is more favorable for sensor efficiency.

To select targets and obtain a list of Type I Seyfert galaxies, we can use the method described in chapter 3 to calculate the redshift of galaxies, by performing a search query in the SIMBAD database.

Not all spectra of type 1 Seyfert galaxies are similar; each case is almost unique, but there are still common lines from a spectrum to another. The key is to see the $H\beta$ line with a good signal-to-noise ratio.

Depending on the magnitude of the target, we can attempt to obtain its spectrum at medium resolution which will allow for greater precision in measuring the FWHM of the $H\beta$ line, but there are hardly any neon lines around this line for calibration. The magnitude limit will quickly be reached and can only apply to the brightest galaxies with a magnitude brighter than 8-9 (It also depends on the optics used).

The use of a low-resolution spectrograph will have the advantage of covering the entire visible spectrum domain; we will be able to see both the various broad and narrow lines of the spectrum and go further in magnitude with a larger selection of galaxies. See also nearby quasars such as 3C273, which have broad lines of hydrogen and other elements.

After the spectrum is processed, the FWHM of the $H\beta$ line is measured and then converted into velocity according to the equation described in Section 8.5 to obtain $V\sin(i)$ from this line. The measured values vary greatly from one galaxy to another, with the following amplitudes:

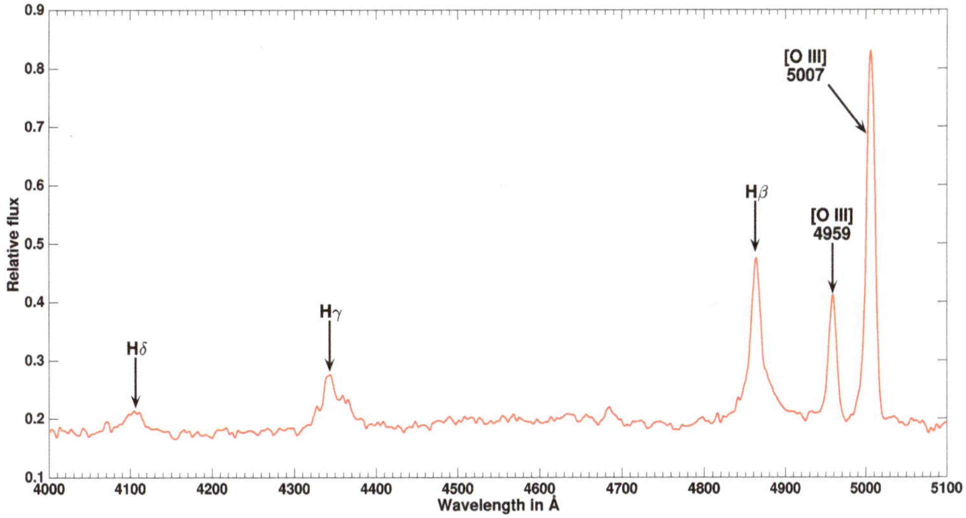

Fig. 8.7.2 — Wide and narrow lines of the galaxy NGC 7469. The hydrogen lines $H\beta$, $H\gamma$ and $H\delta$ are wider than the 2 [O III] lines 4 959 and 5 007 Å. Spectrum obtained from Chile by the 2SPOT team using an Alpy 600 spectrograph, a 305 mm diameter Newtonian telescope at F/4 and an ATIK 414EX CCD camera in 1x1 binning. 4 exposures of 1 200 seconds. On the *x*-axis, the wavelength in ångströms. On the *y*-axis, the relative intensity compared to the continuum normalized to 1.

• Between 1 000 and 10 000 km.s^{-1} for wide lines, as shown in Figure 8.7.2, the rotational velocity is around 1 100 km.s^{-1}.

• Less than 1 000 km.s^{-1} for narrow lines, as shown in Figure 8.7.2, the rotational velocity is around 500 km.s^{-1} on the two [O III] lines.

Having found the rotational velocity of the disk, we can calculate the mass of the black hole if we know the radius between the BLR disk and the black hole according to the equation:

$$M = \frac{RV^2}{G}$$

With:

M: the mass of the black hole in kilograms (kg).

V: the rotational velocities of the disc in km.s^{-1}.

G: the gravitational constant = $6.67428 \cdot 10^{-11}$ $m^3.kg^{-1}.s^{-2}$.

R: the radius or distance between the disk of matter in the BLR zone and the black hole expressed in kilometers. This distance can be obtained according to a relationship that exists between the radius of this disk and the luminosity of the galactic nucleus which can be found on the spectrum continuum at 5 100 Å if it is calibrated in absolute flux.

The mass obtained in kilograms can be converted into solar masses ($M_\odot$) which is more expressive:

$$M_\odot = \frac{M}{1.98892 \cdot 10^{30}}$$

The article "Reverberation measurements for 17 quasars and the size-mass-luminosity relations in active galactic nuclei" by Shai Kaspi *et al.* in *the Astronomical Journal* (ref. 533:631E649, 2000, April 20), indicates how to proceed from a spectrum calibrated in absolute flux (in $erg.cm^{-2}.s^{-1}.Å^{-1}$) using the measured flux of the spectrum at the wavelength of 5 100 Å. We will not go into detail about how to perform this calculation, which requires knowledge of the magnitude of the center of the galaxy that spectroscopic observations should be measured in parallel, but by following this article, you will be able to estimate the mass of the black hole in a type 1 Seyfert galaxy.

8.8 The rotation of the arms of the M77 galaxy

The last type of measurement we will discuss in this chapter involves measuring the rotational velocity of a galaxy from a low- or medium-resolution spectrum.

Let us take the example of a type 2 Seyfert galaxy: M77. This galaxy has the peculiarity of having a fairly large diameter (more than 7 x 6 arc minutes for the visible part) allowing a large area of the spectrograph slit to be covered. However, a focal length greater than 1 500 mm will be required to clearly distinguish the various areas where measurements will be taken. Its relatively low magnitude makes it accessible at medium resolution and will enable measurements to be taken on the arms of the galaxy, which are dimmer than the core.

The outer arms of this galaxy will only be accessible if we take long exposure, however, with exposures of less than an hour, it is already possible to observe the rotation of the brightest arms over a distance of approximately 3 kiloparsecs.

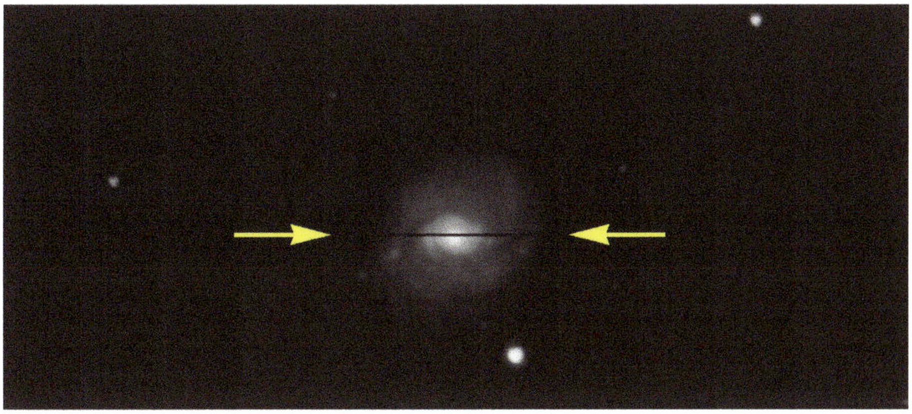

Fig. 8.8.1 — Slit orientation on M77. The two yellow arrows indicate the position of the slit oriented along the major axis of the galaxy. Image taken with the ASI 174 mini autoguiding camera of a LISA spectrograph on a C14 at F/7, with a focal length of 2 500 mm.

If you want to obtain maximum velocities and approximately symmetrical for each arm of the galaxy, the spectrograph must be oriented so that the slit is aligned with the major axis of the galaxy and centred on the nucleus.

The difficulty lies in perfectly centering the galactic core in the slit. The display thresholds for the autoguiding image must be adjusted carefully to make the image of the core as sharp as possible and center it in the slit. Autoguiding will be provided by a star in the field, for example, the one below the slit in Figure 8.8.1.

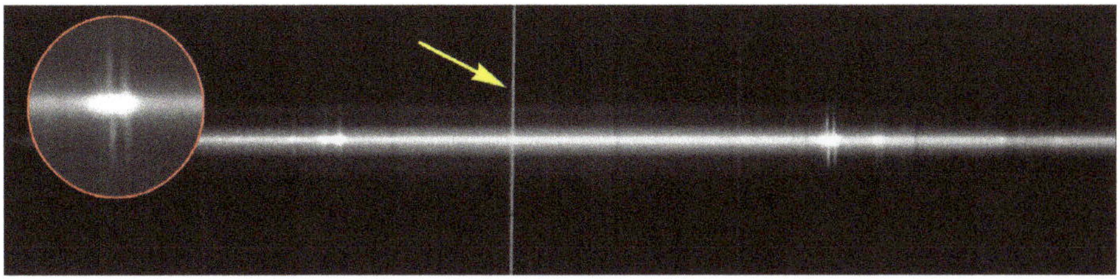

Fig. 8.8.2 — Raw 2D spectrum of M77. The core is very intense. The arm spectrum is visible in the red circle and shows lines shifted on either side of the nucleus. The yellow arrow indicates the atmospheric auroral line [O I] 5 577,35 Å. Spectrum produced with a C14 at F/7, a LISA spectrograph equipped with a 35 μm slit and an ATIK 414Ex CCD camera. Three exposures of 900 seconds.

Be careful not to saturate the spectrum on the $H\alpha$ line because the nucleus is very bright.

The treatment will consist of performing a series of several spectra whose binning height is around 3 pixels, which we will initially take at the center of the galaxy (on the nucleus), and then shift for each spectrum, the Y value of the center of the 3-pixel binning area, towards the top of the 2D spectrum, then towards the bottom.

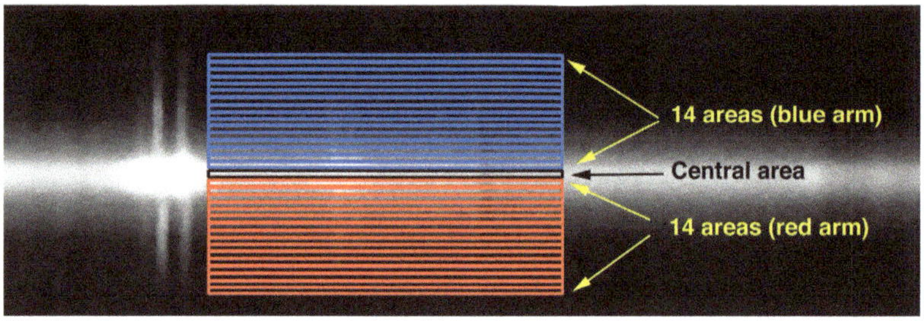

Fig. 8.8.3 — M77 binning areas. Definition of the various binning areas of the spectrum, with the central area corresponding to the galaxy's core shown in black, in red, the successive binning areas of the galactic arm moving away from the observer, In blue, the binning areas closest to the observer.

We obtain a series of about thirty spectra shifted by 3 pixels and which covers the visible arms of the galaxy. The treatment is done without removing the background, because the auroral atmospheric line [O I] 5 577.35 Å is very useful for checking the wavelength calibration of each individual spectrum.

If there is a discrepancy between the measured wavelength and the expected theoretical value, the entire spectrum will need to be shifted by this discrepancy. For example, if you measure the position of this line at 5 577.50 Å, is that it is shifted by +0.15 Å from the theoretical value.

ISIS allows you to make this correction using the function "**Shift**" from the tab "**5-Profile**" where the value is indicated in ångströms of the desired offset, as shown in Figure 8.8.4.

It is imperative that all spectra be calibrated in wavelength as accurately as possible because the velocity measurements we are going to take will be very close to the instrumental error of the spectrograph especially if you are using low resolution.

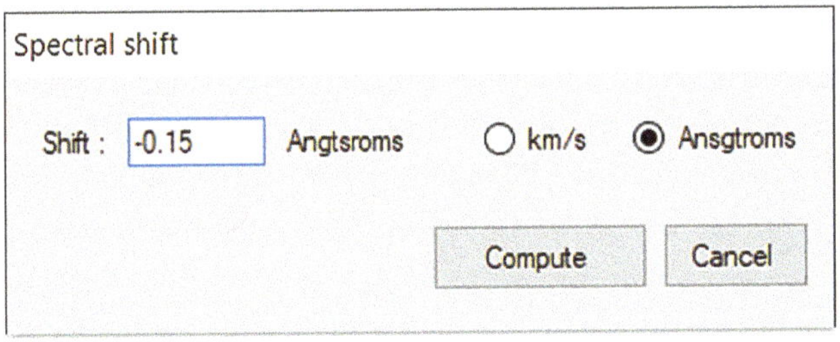

Fig. 8.8.4 — Window "Spectral Shift". From the ISIS software.

Each spectrum can also be corrected for heliocentric and galactocentric velocity although the induced error remains below the instrumental error of the spectrograph if the resolution is low but can become significant if you are using medium resolution. You must then identify in each spectrum 3 to 4 lines on which the observed wavelengths will be measured.

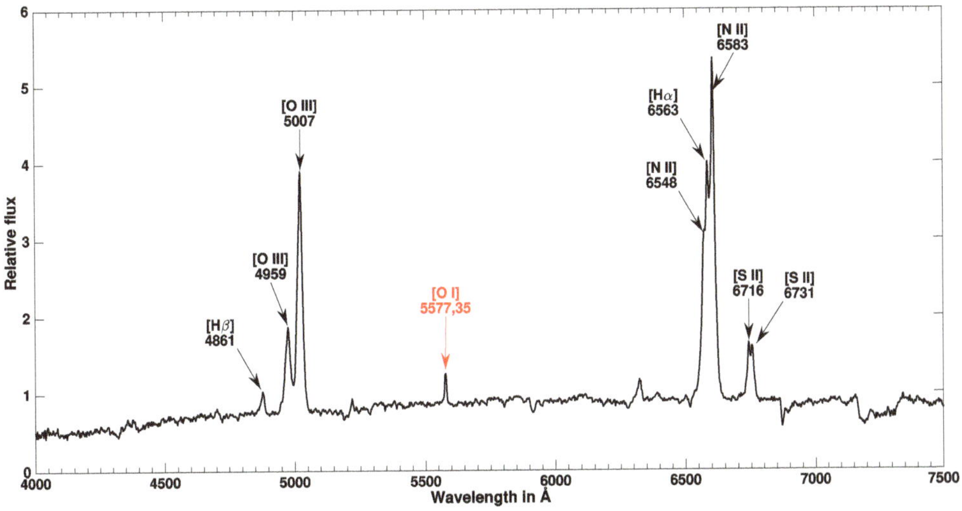

Fig. 8.8.5 — Spectrum of M77. Identification of lines that can be used to measure spectral shift. The line identified in red is the atmospheric auroral line which is only used to verify correct wavelength calibration. On the x-axis, the wavelength in ångströms. On the y-axis, the relative intensity compared to the continuum normalized to 1.

For each identified line, the redshift is measured (see section 3.4 for the measurement method to be adopted). Avoid taking mixed rays whose uncertainty of measurement will be larger, like the line [N II] 6 548 Å, which is too close to the $H\alpha$ line, However, the two lines [S II] can be used because the two peaks are clearly separated. The results are recorded in a spreadsheet and the average redshift of the measured lines is calculated for each spectrum. This procedure is repeated for all spectra in the series. The average redshift z is converted to velocity using the following equation:

$$V_z = cz$$

Where c is the speed of light (299 792.458 km.s^{-1}) and V_z the recession velocity along the line of sight in km.s^{-1}.

The velocity of each spectrum naturally includes the value of the galaxy's recession velocity which must then be subtracted to obtain only the velocity of each spectrum relative to the galaxy's core. The redshift of M77 according to the SIMBAD database is z = 0.003 483 72, with a recession velocity of 1 044.39 km.s^{-1}. The measured velocity of the nucleus can also be reduced to zero and this correction applied to all other spectra.

This produces a list of values in the spreadsheet ranging from -160 to +160 km.s^{-1} presentable in graph form.

Fig. 8.8.6 — Velocity graph of the M77 galaxy. Measurements of the velocities of the 27 spectra that were taken on this galaxy every 3 pixels. On the x-axis, the distance from the galactic core measured in kiloparsecs. On the y-axis, the rotational velocity in km.s^{-1}.

8.8.1 Conversion of distance in pixels / distance in parsecs

This conversion requires knowledge of the sampling rate of the autoguiding sensor image and the magnification factor between this image and the spectrum.

1) Measurement of the autoguiding image sampling

So we have the value of the velocity (in $km.s^{-1}$) for a given position measured in pixels in the spectrograph slit. This position in pixels must be converted into angular distance from the center of the galaxy.

This distance is expressed in kiloparsecs (see the chapter 3 for equivalent values in astronomical units or light years).

First, you need to know the sampling rate of the image produced on the autoguiding sensor in arc seconds per pixel. This can be easily determined by performing an astrometric field survey of an image taken with the autoguiding camera.

Some astronomy software programs allow you to take this measurement. This empirical relationship can also be used to calculate this sampling:

$$E = 206\left(\frac{P}{F}\right)$$

With:

E: the value of the image sampling in arc seconds per pixel.
P: the size of a pixel in μm.
F: the focal length of the telescope or refractor in mm.

A very useful online tool also provides this information with greater accuracy:

https://nova.astrometry.net/

where the result indicates the scale of the image (*pixel scale*), in arc seconds per pixel. This value must then be converted to radians using the following formula:

$$1'' \frac{\pi}{(180 \times 3\,600)} = 4.8481 \cdot 10^{-6} \text{ rad}$$

The distance to M77 in parsecs is calculated from its recession velocity and the Hubble constant (see section 3.4). With a value for the Hubble constant of $H_0 = 74.03$

$km.s^{-1}.Mpc^{-1}$ and a recession velocity of 1 044.39 $km.s^{-1}$, we obtain a distance D in megaparsecs of:

$$D = \frac{V_z}{H_0} = \frac{1\,044.39}{74.03} = 14.107 \text{ Mpc}$$

or 14 107 kpc (kiloparsecs).

To find the angular separation in kiloparsecs of a pixel on the image, we multiply the distance in kiloparsecs of M77 by sampling E of a pixel in radians.

In the case of observing M77 with a C14 at F/7 and an ASI 174 mini binned 2x2, we have a sampling E = 1.02 arc seconds per pixel, which are converted to radians using the equation described above. The value of a sampling E in radians is obtained as:

$$E_{(rad)} = 1.02 \frac{\pi}{(180 \times 3\,600)} = 4.945 \cdot 10^{-6} \text{ rad}$$

and the angular value in kiloparsecs of a pixel on the autoguiding image

$$\text{M77 distance(kpc)} \times E_{(rad)} = 14\,107 \times 4.945 \cdot 10^{-6} = 0.07 \text{ kpc/pixel}$$

Each pixel in the autoguiding sensor image corresponds to 0.07 kiloparsecs.

2) Spectrum sampling measurement

Please note! The previously calculated angular value is only valid for the autoguiding sensor because according to the spectrograph model, the image passing through the slit may undergo a dimensional factor which will modify the magnification ratio of the spectrum produced on the spectral camera.

It is therefore necessary to calculate this magnification or reduction ratio between the autoguiding image and the 2D spectrum. Here is an example of a spectrum obtained for M77 using a LISA spectrograph.

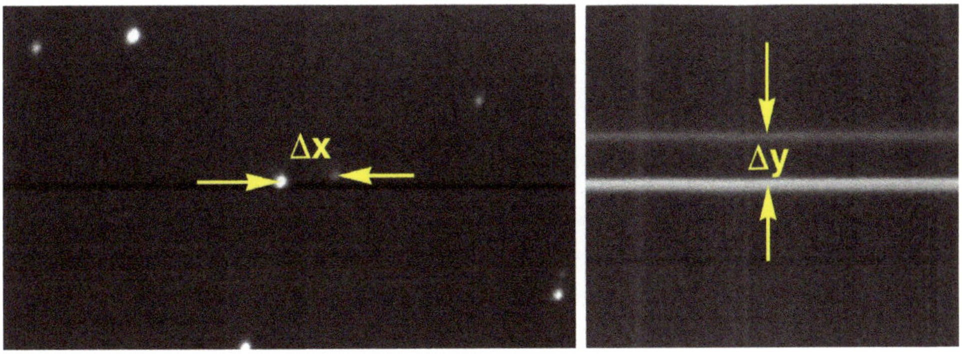

Fig. 8.8.7 — Measure the dimension ratio. On the left is the image from the autoguiding sensor, and on the right is the image from the spectral camera.

This factor can be measured by placing two stars in the slit of the spectrograph. The distance between the two stars is measured both on the autoguiding sensor (Δ_x) and on the 2D spectrum (Δ_y).

In the case of the configuration used and the example in the figure 8.8.7, We measure $\Delta_x = 30$ pixels (autoguiding camera) and $\Delta_y = 33$ pixels (spectral camera). We calculate the ratio between the two:

$$Fg = \frac{\Delta_x}{\Delta_y} = \frac{33}{30} = 1.1$$

The image on the spectral sensor is 1.1 times larger than on the autoguiding sensor.

This magnification factor Fg is then applied of x1.1 on the angular value per pixel of 2D spectra. We now know the factor to apply between the image from the autoguiding sensor (or an image taken with another instrument) and that of the 2D spectrum. Comparison of spectral data with an image can now be achieved with the same angular scale.

The velocities are, of course, measured according to the observer's angle of view and do not take into account the inclination (i) of the galaxy. It is therefore the projected velocities, not the real velocities, that need to be corrected by estimating the inclination of the galaxy.

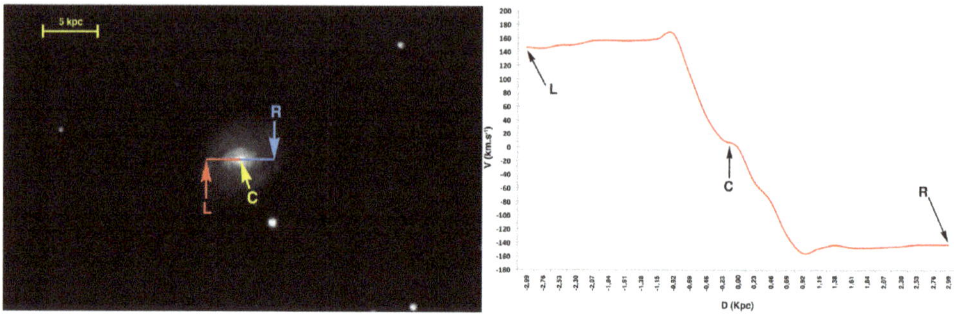

Fig. 8.8.8 — The velocities of M77 according to distance from the core. Correspondence of angular differences between the graph of measured velocities (km.s^{-1}) and the image of M77 calibrated in kiloparsecs.

To do this, we can model an ellipse around the visible part and thus estimate its inclination, but the exercise remains difficult to carry out because we can only see a tiny part of the galaxy in the visible spectrum domain and this extends much further into other wavelengths (For example, the 21 cm hydrogen line, which is only accessible in radio astronomy). However, the non-visible extensions are in the same plane, except for a few rare cases of "disturbed" galaxies.

Despite the small observed radius of 3 kiloparsecs, we were able to measure these velocity variations.

8.8.2 A clue to dark matter?

We can see that velocities evolutions depending on the distance from the center in each arm of the galaxy, are not Keplerian and tend to remain at a constant plateau as soon as one moves away from the center of the galaxy beyond a distance of one kiloparsec. If the rotation were Keplerian, the rotational velocity should decrease, but this is not the case!

So there is "something" that keeps these velocities at an almost constant level in the galaxy's arms.

The most likely hypothesis is the presence of invisible dark matter, which encompasses the galaxy and contributes significantly to its total mass. With these amateur measurements, we can confirm the presence of dark matter around a galaxy.

8.9 Bibliography

- Astronomie, Astrophysique, Introduction 4th edition by Agnès Acker, Dunod, 2005. ISBN 2 10 005462 7.

- Atomic Astrophysics and Spectroscopy by Anil K. Pradhan and Sultana N. Nahar, Cambridge University Press, 2011. ISBN 978-1-107-48358-3.

- Classical Be Stars: Rapidly Rotating B Stars with Viscous Keplerian Decretion Disks by Thomas Rivinius, Alex C. Carciofi and Christophe Martayan, Astronomy and Astrophysics, 2013, DOI: 10.1007/s00159-013-0069-0.

- Accretion disks in Seyfert nuclei: broad line profiles and asymmetries by E. van Groningen, Astronomy and Astrophysics, 126, 363-371, 1983.

- Profiles and profile ratios in Seyfert 1 galaxies by D. Michael Crenshaw, The Astronomical Journal Supplement series 62: 821-838, 1986 december.

- Stellar V sin i and optical emission line widths in Be stars by Reinhard W. Hanuschik, Astrophysics and Space Science (ISSN 0004-640X), vol. 161, no. 1, Nov. 1989.

- Non-coherent scattering in vertically extended Be star disks winebottle-type emission-line profiles by W. Hummel and J. Dachs, Astronomy and Astrophysics 262, L17-L20, 1992.

- A study of high-resolution emission-line profiles for Be stars by J. Dachs, W. Hummel and R.W. Hanuschik, Astronomy and Astrophysics Supplement Series 95, 437-460, 1992.

- Non-Axisymmetric Be star circumstellar disks by W. Hummel and M. Vrancken, Astronomy and Astrophysics 302, 751-764, 1995.

- Non-Keplerian rotation in the nucleus of NGC 1068: evidence for a massive accretion disk by G. Lodato and G. Bertin, Astronomy and Astrophysics 398, 517-524, 2003. DOI: 10.1051/0004-6361:20021672.

- Fourier method in the determination of rotational velocities in OB stars by S. Simon-Dıaz and A. Herrero, Astronomy and Astrophysics 468, 1063-1073, 2007. DOI: 10.1051/0004-6361:20066060.

- High S/N Echelle Spectroscopy in Young Stellar Groups II. Rotational Velocities of

Early-Type Stars in Sco OB2 by A.G.A. Brown and W. Verschueren, Astronomy and Astrophysics, vol. 319, p. 811-838, 1997.

- Reverberation measurements for 17 quasars and the size-mass-luminosity relations in active galactic nuclei by Shai Kaspi, Paul S. Smith, Hagai Netzer, Dan Maoz, Buell T. Jannuzi et Uriel Giveon, The Astrophysical Journal, vol. 533, no 2, 2000 ApJ 533 631. DOI: 10.1086/308704.

- A study of high-resolution emission-line profiles for Be stars by J. Dachs, W. Hummel and R.W. Hanuschik, Astronomy and Astrophysics Supplement Series 95, 437-460, 1992.

- Rotational velocities of A-type stars I. Measurement of V sin i in the southern hemisphere by F. Royer, M. Gerbaldi, R. Faraggiana and A. E. Gomez, Astronomy and Astrophysics 381, 105121, 2002. DOI: 10.1051/0004-6361:20011422.

- The shape of broad-line profiles in active galactic nuclei by W. Kollatschny and M. Zetzl, Astronomy and Astrophysics 549, A100, 2013. DOI: 10.1051/0004-6361/201219411.

- Accretion disk wind as explanation for the broad-line region structure in NGC5548 by W. Kollatschny and M. Zetzl, Astronomy and Astrophysics, 551, L6, 2013. DOI: 10.1051/0004-6361/201220923.

- Critical study of the distribution of rotational velocities of Be stars II: Differential rotation and some hidden effects interfering with the interpretation of the V sin i parameter by J. Zorec and al., Astronomy and Astrophysics, 602, A83, 2017. DOI: 10.1051/0004-6361/201628761.

- The projected rotational velocity distribution of a sample of OB stars from a calibration based on synthetic He I lines by Simone Daflon, Katia Cunha, Francisco X. de Araujo, Sidney Wolff and Norbert Przybilla, The Astronomical Journal, 134:1570Y1578, 2007 October.

- The large scale trend of rotation curves in the spiral galaxies NGC 1068 and NGC 3310 by G. Galletta and E. Recillas-Cruz, Astronomy and Astrophysics, 112, 361-365, 1982.

- Rotational velocities of B stars by Helmut A. Abt, Hugo Levato and Monica Grosso, The Astrophysical Journal, 573:359-365, 2002 July 1.

- Be-star rotation: how close to critical? by R. H. D. Townsend, S. P. Owocki and I. D. Howarth, MNRAS, 350, 189-195, 2004.

- Disc tearing in a Be star. predicted 3D observations by M. W. Suffak, C. E. Jones and A. C. Carciofi, MNRAS 527, 7515-7522, 2024.

- Spectroscopic Line Modeling of the Fastest Rotating O-type Stars by Katherine Shepard, Douglas R. Gies, Lex Kaper, and Alex De Koter, The Astrophysical Journal, 931:35 (22pp), 2022 May 20.

9. Time series in spectroscopy

This chapter describes the methods that can be used to produce temporal spectral series (Time series).

Some objects undergo rapid change over a period of a few hours or even a few minutes.

In order to monitor these variations, a continuous sequence is required. This may need to be rather long.

Changes can only be detected with the highest possible signal to noise ratio, which is not necessarily obvious, a condition that is not always easy to monitor. A compromise must be found between sampling time and unit exposure time which depends on the magnitude of the object being studied, but also on the spectral resolution applied.

Figure 9.0.1 shows a series of 12 spectra of RR Lyr star taken every 10 minutes, for a period of 2 hours.

Changes can be observed in the profiles of lines D1 and D2 (the sodium lines at 5 890 and 5 896 Å) with a doubling of these two lines due to the fact that part of the stellar atmosphere rises and therefore has a different velocity from that which remains at rest. This is the Schwarzschild effect.

What is interesting and only appears on a single 10 minute exposure is the emission line of He I at 5 876 Å (circled in yellow on the graph).

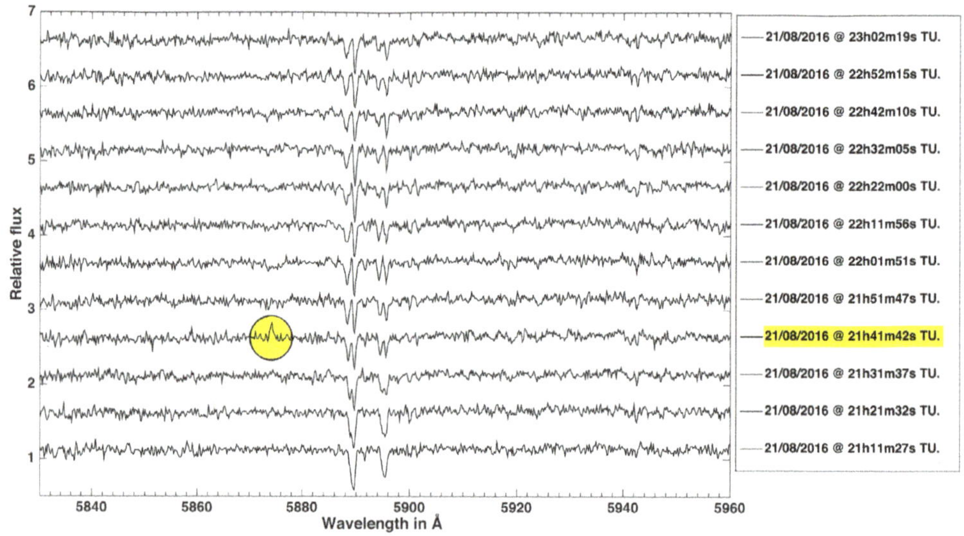

Fig. 9.0.1 — Spectral time series of the RR Lyr star. Centred on sodium lines D1 and D2 and He I 5 876 Å. Produced using an eShel spectrograph ($R = 11\ 000$). Exposure time of 600 seconds over a period of 2 hours. 406 mm diameter Ritchey-Chrétien telescope and an ATIK 460EX CCD camera. On the x-axis, the wavelength in angstroms. On the y-axis, relative intensity based on the continuum normalized to 1.

This occurrence can be dated with an error bar of ± 10 minutes, this accuracy could have been greater if had the spectral exposure time been below 10 minute, which is not always feasible, whilst maintaining a good signal-to-noise ratio.

It is thanks to this series of short exposures that we can highlight these very brief phenomena that occur during certain cycles of this star.

9.1 Dynamic 2D graphics

Another way of presenting the evolution of a spectrum over time is to create a dynamic 2D graph in color or grayscale with wavelength on the x-axis and dates on the y-axis.

Colour shades or shades of grey emphasise variations in spectral profile intensity according to the scale indicated below the graph (the z-axis of the graph).

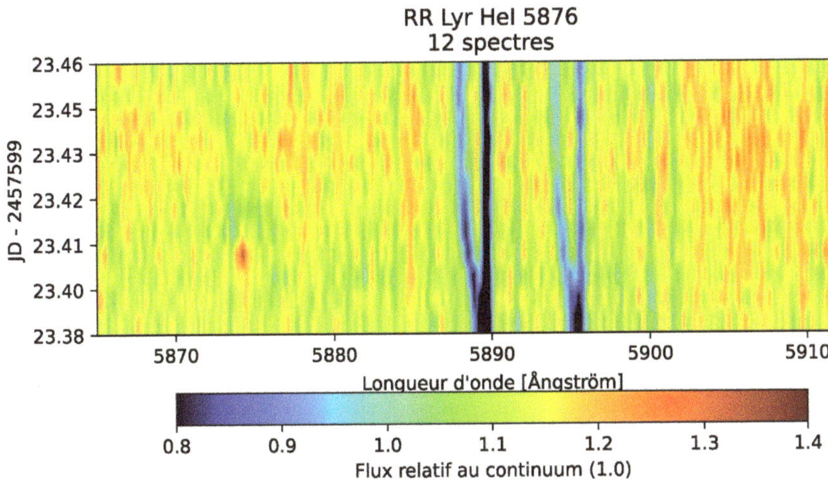

Fig. 9.1.1 — Dynamic 2D graph of RR Lyr. Centred on sodium lines D1 and D2 and He I 5 876 Å. Created from the same data as Figure 9.0.1. On the *x*-axis, the wavelength in angstroms. On the *y*-axis, the fraction of the Julian date minus 2 457 599.

In the example shown in Figure 9.1.1, the continuum has been normalized to 1, Blue/green colors represent absorption lines (intensity <1), whilst yellow/orange/red colors represent emission lines (intensity >1).

The 12 spectra in this sequence have been arranged vertically and in chronological order (the graph is read chronologically from bottom to top).

The emission of He I 5 876 Å line is clearly visible on this type of graph: it is the isolated red dot on the left. Similarly, the evolution and splitting of the D1 and D2 bands are more visible.

ISIS or Visual Spec software offers the ability to create dynamic 2D graphs. A Python script can also be used to generate this type of graph.

Figure 9.1.2 shows an example of a script written by Guillaume Bertrand. It has been modified for this specific example, but can easily be adapted to suit your own needs.

Some instructions before running this script:

The folder containing the script must include a folder (here named "dataRRLyr1") containing the spectra already processed. They begin with the root name: @pro, in this example: @pro38-1.fit, @pro38-2.fit, @pro38-3.fit..... @pro38-12.fit.

```python
import os
import copy
import glob
import warnings
import matplotlib.pyplot as plt
import matplotlib.ticker as mticker
import numpy as np
import astropy.wcs as fitswcs
from astropy.io import fits
from astropy.time import Time
import astropy.units as u
from specutils import Spectrum1D
def find_and_load_spectra(path, pattern):
    results= []
    for filename in sorted(glob.glob(os.path.join(path, pattern))):
        spectrum = fits.open(filename)
        header = spectrum[0].header
        with warnings.catch_warnings():
            warnings.simplefilter('ignore')
            wcs_data = fitswcs.WCS(header={'CDELT1': round(header['CDELT1'],4),
            'CRVAL1': header['CRVAL1'], 'CUNIT1': 'Angstrom', 'CTYPE1': 'WAVE',
                                       'CRPIX1': header['CRPIX1']})
            spec1D = Spectrum1D(flux= spectrum[0].data * u.Jy,  wcs=wcs_data)
            jd = Time(header['DATE-OBS'], format='isot', scale='utc').jd
            results.append((spec1D, jd))
    results = np.array(results)
    results = results[results[:,1].argsort()]
    return results
def plotSpec2DFlux(spectra, title="", subtitle="", savefig=None, sub_jd=2450000,
                   cmap="gist_gray", fig_size=(7, 4)):
    min_wvl, max_wvl = spectra[0][0].wavelength.min().value, spectra[0][0].wavelength.max().value
    min_jd, max_jd = min(spectra[:,1]), max(spectra[:,1])
    spec2d = np.zeros((len(spectra[0][0].wavelength), len(spectra)))
    for i, (s, jd) in enumerate(spectra):
        spec2d[:, i] = s.flux
    fig, ax = plt.subplots(figsize=fig_size)
    plt.imshow(np.rot90(spec2d),  extent=[min_wvl, max_wvl, 0, len(spectra)], aspect='auto', cmap=cmap)
    ax.set_xlabel('Longueur d\'onde [Ångström]',fontdict=None, labelpad=None)
    ax.set_title("%s\n%s" % (title, subtitle))
    ticks_loc = ax.get_yticks().tolist()
    ax.yaxis.set_major_locator(mticker.FixedLocator(ticks_loc))
    ax.set_yticklabels(['{:.2f}'.format(a) for a in np.linspace(min_jd-sub_jd, max_jd-sub_jd,
    len(ax.get_yticks()), dtype=float)])
    ax.set_ylabel(f'JD - {sub_jd}', fontdict=None, labelpad=None)
    plt.tight_layout(pad=1, w_pad=1, h_pad=1)
    plt.clim(0.8,1.4)
    plt.colorbar(label='Flux relatif au continuum (1.0)', location='bottom', shrink=0.8)
    plt.xlim(5865,5915)
    if savefig:
        plt.savefig(savefig)
    plt.show()
if __name__ == '__main__':
    spectra = find_and_load_spectra("dataRRLyr1","@pro*.fit")
    plotSpec2DFlux(spectra,
                   title="RR Lyr HeI 5876",
                   subtitle=f"{len(spectra)} spectres",
                   sub_jd=2457599,
                   cmap="turbo",#autumn plasma, ocean, gnuplot,| gist_stern
                   savefig="Graphe2D.pdf")
```

Fig. 9.1.2 — Python script for dynamic 2D graphs.

The .fits header of the spectra must include the keyword "DATE-OBS" so that the script can retrieve the observation date in the format 2024-01-18T23-13-25 (year-month-day-hour-minute-second). This essential information is already included in the fits header if the processing was carried out using software such as Demetra, ISIS, or Spec INTI.

The folder name and the root name of the spectra can be changed on line 53.

When you run the script for the first time, line 48, which sets the wavelength limits of the graph, must be deleted. Then, depending on the desired spectral interval for your graph, you can redefine the limits by adding an amended line of code (e.g. plt.xlim(6500, 6700)).

Line 55 allows you to give the graph a title.

Line 56 indicates the number of spectra that were used to generate the graph.

Line 58 defines the color palette used, here the one named "turbo". There are several. Comments on the same line have been added with the names of other possible palettes. Others are proposed in the documentation of the Python library, matplotlib:

https://matplotlib.org/stable/users/explain/colors/colormaps.html

Line 59 allows you to save the graph in the format of your choice: .pdf, .png or .jpg.

9.2 BW Vul, a short-period pulsating star

Here is an example of a time series, easy to capture, of the pulsating star BW Vul, whose period is approximately 4 hours and 49 minutes, which allows the observation to be carried out in a single night.

Information about this star:

BW Vul (HD 199140), magnitude $V = 6.54$, spectral type: B2IIIv

RA: 20h 54' 22.39" (J2000)
DEC: +28° 31' 19.19" (J2000)

It is a beta Cephei type star, a category of star that has the distinctive feature of having regular and fairly rapid pulsation periods ranging from 0.1 to 0.3 days. Beta Cephei was the first pulsating star discovered in 1902 by American astrophysicist Edwin Brant Frost (1866-1935).

Frost discovered that the pulsation of the star is the cause of the variations in velocity measured on the spectra. With each cycle, the star's photosphere "swells" and then returns to its initial position before inflating again, which causes these periodic variations in velocity.

Variable stars are grouped by type according to their light curves, their period, their own physical characteristics among other things. The generic name of these groups is often that of the star used for classification. So we have the following types: RR Lyr, Mirae, T Tauri, Cepheids (several subtypes) and many others.

These are stars that show fairly significant variations in radial velocity that can be measured with amateur equipment. BW Vul is the pulsating star with the greatest amplitude, around 200 $km.s^{-1}$ associated with a variation in brightness of ±0.11 magnitude.

To clearly highlight the evolutions in the spectrum of this star, it is necessary to use a medium or high resolution spectrograph ($R > 5\ 000$) such as Star'Ex HR, LHIRES III, or a multi-order Echelle type spectrograph such as an eShel or BACHES.

It is important to determine the best compromise between the unit time exposure and the number of spectra to be produced to obtain adequate temporal sampling and highlight the phenomenon, while maintaining a good signal-to-noise ratio.

With an average magnitude of approximately $V = 6.54$, a single exposure time of 600 seconds can be used on most instrument configurations. Of course, this value must be adjusted according to the spectrograph and optics used.

The higher the sampling rate, the better the sampling will be at showing small temporal variations in the profile, but at the expense of a lower signal-to-noise ratio. It is sometimes a difficult choice!

Several lines can be used to show variations in velocity, such as hydrogen lines or helium lines.

The observation consists of producing a series of spectra covering at least one pulsation period of the star. In this example, we used an eShel spectrograph with a 406 mm diameter Ritchey-Chretien telescope. 31 spectra of 600 seconds each were taken to cover slightly more than the star's period. They are treated individually using the same protocol.

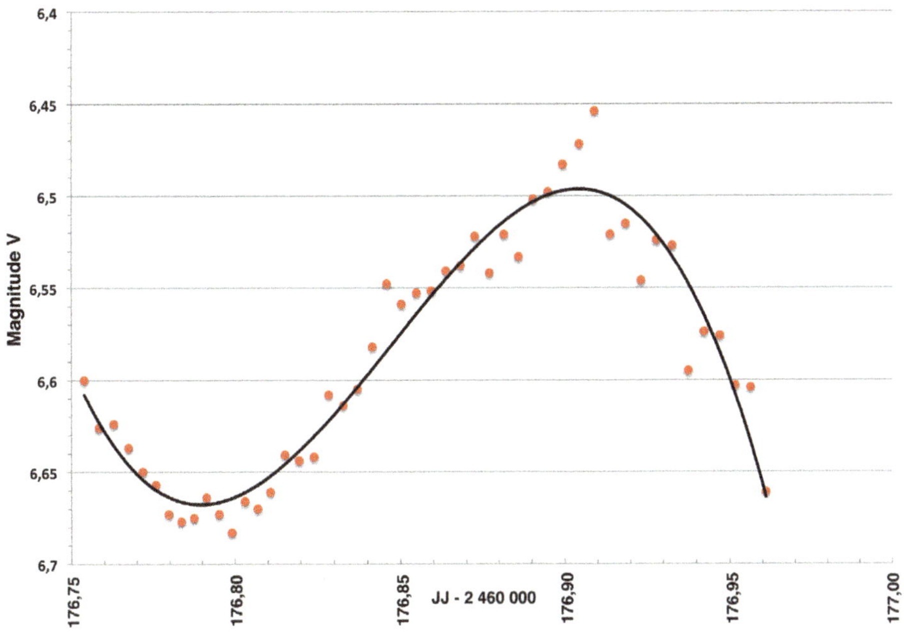

Fig. 9.2.1 — Variation in the *V* magnitude of BW Vul over one period. Photometric data extracted from the AAVSO database: www.aavso.org. On the *x*-axis is the Julian day, on the *y*-axis is the star's *V* magnitude.

9.2.1 Processing a series with ISIS

ISIS has a function that allows you to process an entire series of spectra at once. Here's how to proceed:

In the tab "**Settings**", you must check "**no**" under "**Erase intermediate files**".

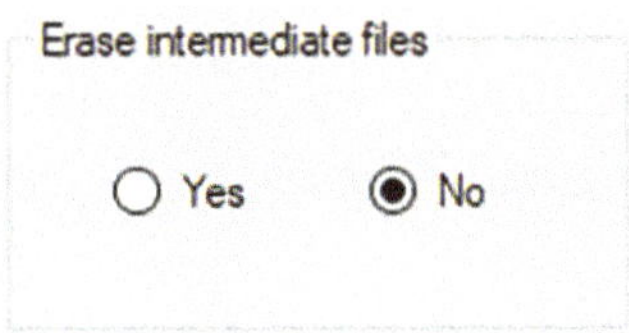

Fig. 9.2.2 — Window "Erase intermediate files" from the tab "Settings". From the ISIS software.

This preserves the files containing the unit spectra "intermediaries" generated during processing, otherwise they are stacked to obtain the final spectrum and then deleted. Their generic name is "@pro" followed by the spectrum number and the .fits extension, They are located in the working directory. These will be used to generate the dynamic 2D graph.

If you use the script provided in the section 9.1, Then you need to gather all these spectra into a single folder, which you place in the Python script folder.

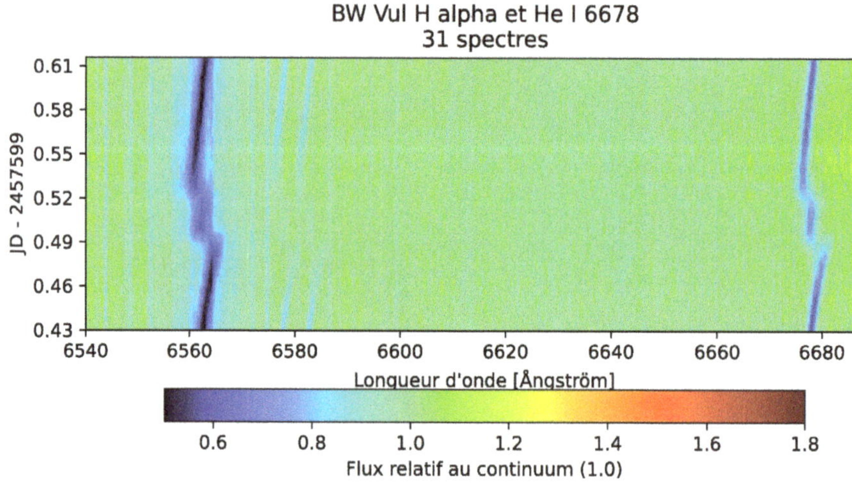

Fig. 9.2.3 — Dynamic 2D graph of BW Vul. Range covering the $H\alpha$ line (left) and He I 6 678 Å (right). 31 spectra of 600 seconds each, taken with an eShel spectrograph (R = 11 000) and a 406 mm diameter Ritchey-Chretien telescope. On the x-axis, the wavelength in angstroms. On the y-axis, the fraction of the Julian date minus 2 457 599.

The graph shows straight vertical lines on either side of the $H\alpha$ line, These are atmospheric lines; they are not of stellar origin but terrestrial, which is why they do not vary!

The distorted lines are produced by the star during the pulsation period and exhibit velocities variations of around 200 km/h.s^{-1}.

By using only the helium I line at 6 678 Å, which has no atmospheric lines in its vicinity (on the right side of Figure 9.2.3), we obtain a more readable graph that allows us to highlight the physical evolution of the star during a given period in greater detail.

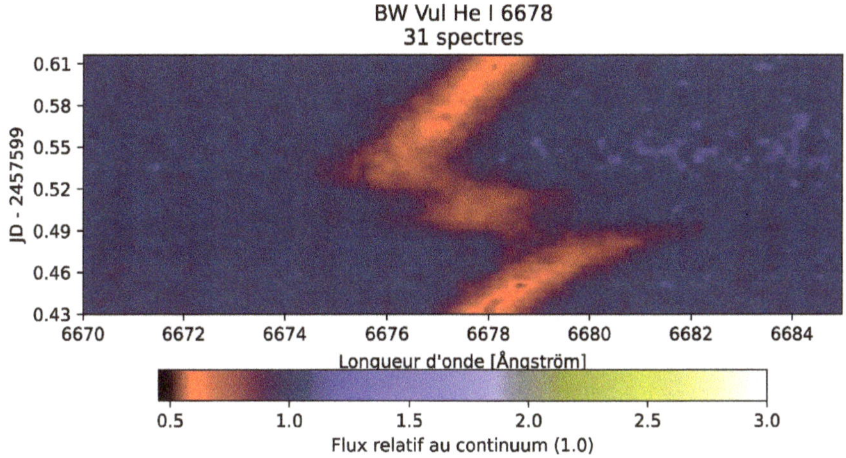

Fig. 9.2.4 — Dynamic 2D graph of BW Vul. Centred on the He I line at 6 678 Å. Obtained from the data used for Figure 9.2.3. On the *x*-axis, the wavelength in angstroms. On the *y*-axis, the fraction of the Julian date minus 2 457 599.

The graph in Figure 9.2.4 shows two breaks in the profile of the helium I 6 678 Å line, which are the signature of two shock waves occurring in the star's atmosphere during a pulsation period.

These are two layers of the star's atmosphere that move in opposite directions. One rises and the other falls and their collision creates the shock waves seen in Figure 9.2.4.

The horizontal red section around 0.49 (Julian day minus 2 457 599 for greater clarity) is considered to be the "rest" phase of the star's atmosphere where shock waves are not present. That part is often called "*standstill*" (quiet phase) in publications. It is located between the two shock waves. Other spectral lines also show the same type of 2D dynamic profile.

You can try to create time series for other pulsating stars but since the velocity range is smaller, the measurements will be harder to detect. Some pulsating stars also are fairly dim stars, which makes them difficult to access using medium or high resolution spectroscopy with amateur equipment.

To obtain a list of this type of star, you can run a query in the SIMBAD database as described in the section 3.2 using the command line:

otype = "bcepV*" & **Vmag<7**

In return, you will receive a list of approximately 60 pulsating stars of the Beta Cephei type whose magnitude V is less than 7. Not all of these stars will necessarily be within reach of amateur spectrographs for velocity measurements but the brightest ones may be attempted with single exposures shorter than 600 seconds. For example, Beta Cephei has a velocity amplitude of $\pm$ 17 km.s^{-1} and a period P = 0.19 day, or 4.56 hours with a magnitude V that varies from 3.16 to 3.27.

9.3 Study RR Lyr

We mentioned at the beginning of this chapter that the star RR Lyr showed emission lines for a few minutes during its pulsation period: P = 0.566782 day which is just over 13.6 hours. It is difficult to obtain a complete period during one night of observation, It is therefore necessary to choose the right moment to record the spectra that show the most significant and interesting developments.

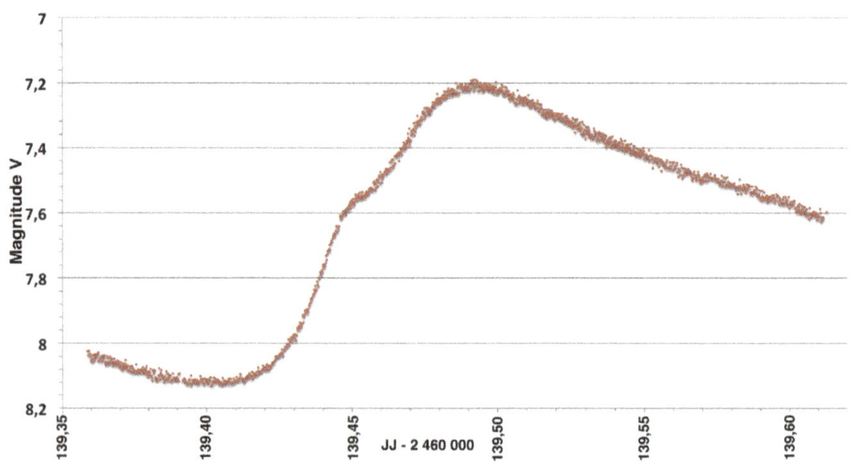

Fig. 9.3.1 — Light curve of RR Lyr. According to photometric data in the V band from the AAVSO database: www.aavso.org

The photometric maximum of RR Lyr does not correspond to the spectroscopic maximum. The spectroscopic maximum, i.e., the moment when the $H\alpha$ line transitions to emission, occurs at phase 0.91 of its period, approximately 1 hour and 13 minutes before the photometric maximum (at maximum brightness). To see the evolution of the spectral profile close to this moment, you must begin observing RR Lyr at least two hours before its photometric maximum and continue observations for at least 1 hour afterwards. It is therefore necessary to know when this event occurs. Here are two methods:

With photometric data from the AAVSO

Data from the AAVSO website can be used: www.aavso.org

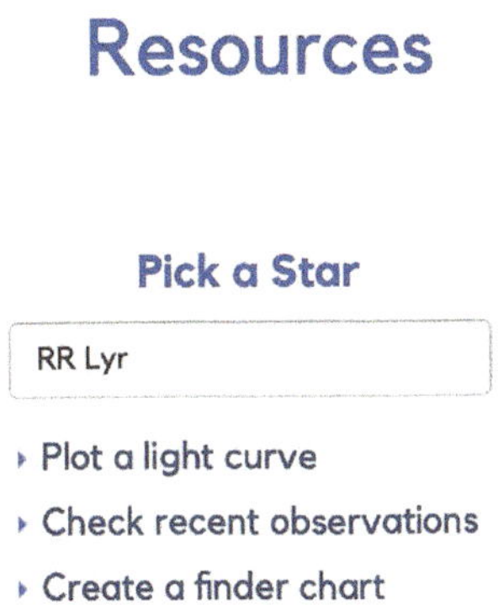

Fig. 9.3.2 — Area "Resources" from the AAVSO website. Available on the website's home page: www.aavso.org

On the home page, look for the area "**Resources**" and enter the name of the star, then click on "**Plot a light curve**". In return, you will receive the photometric data for this star and you can view the last period measured with the date of the last photometric maximum. Preferably use photometric band V. Knowing the period $P = 0.566782$ day, you can determine the date and time of the next maximum that interests you.

Using an online ephemeris generator

There are many websites for generating ephemerides, an internet search will help you find some. As the presentations are very diverse, it is difficult to recommend a particular website, especially since its future existence is not guaranteed.

Now that you have determined the date and time of the start of observation, the exposure time for each spectrum must be estimated. As indicated for BW Vul, this unit time will depend on your spectrograph and optics. You will need to find the best possible compromise between exposure time and the signal-to-noise ratio of the spectrum obtained but remaining within values that allow observation of the $H\alpha$ emission line or even that of helium at 5 876 Å (rarer: this occurrence depends on the secondary phase ± 39 days: Blazhko effect) on an exposure of less than 600 seconds.

You can then take a series of spectra of RR Lyr using the following coordinates:

RR Lyr (HIP 95497)

RA: 19h 25' 27.91" (J2000).

DEC: +42° 47' 03.69" (J2000).

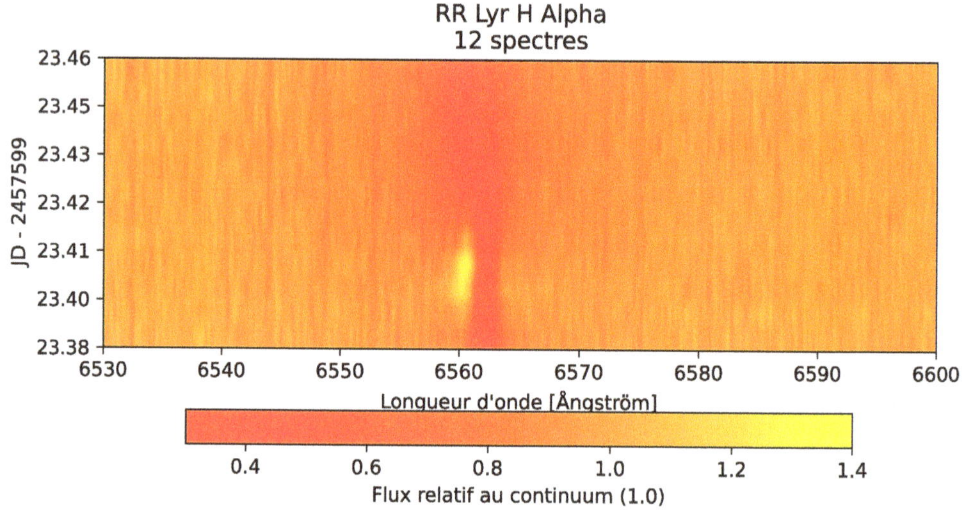

Fig. 9.3.3 — Dynamic 2D graph of RR Lyr. Centered on the $H\alpha$ line. 12 single exposures of 600 seconds each, taken with an eShel spectrograph and a 406 mm diameter Ritchey-Chretien telescope. On the x-axis, the wavelength in angstroms. On the y-axis, the fraction of the Julian date minus 2 457 599.

On the graph in Figure 9.3.3, we observe the emission of the $H\alpha$ line over a period of approximately 14 minutes (in yellow on the graph) and the doubling of this line. From one cycle to another, the intensity of the emission lines ($H\alpha$ and/or He I 5876) will be different because another phenomenon overlaps this pulse cycle found in RR Lyr-type stars, being the Blazhko effect, named after the Russian astronomer Sergey Blazhko (1870-1956).

Blazhko discovered for another RR Lyr-type star (RW Dra) that a secondary variation in the pulse affects both the period and the amplitude over several cycles.

This secondary period due to the Blazhko effect on RR Lyr is approximately 39 days. It can be highlighted by performing spectral series on RR Lyr over several periods.

Please note that this effect varies, it would appear that a third, much longer period (years) also exists and overlaps with the previous two.

9.4 Study the flickering of spectral lines

Some stars have spectral profiles that evolve stochastically over time and often for short periods ranging from a few minutes to a few hours. We will discover how to observe this twinkling (or *flickering*) with a concrete example on the star WR 11. The Wolf-Rayet type stars (WR) were discovered in 1867 by French astronomers Charles Wolf (1827-1918) and Georges Rayet (1839-1906) who observed broad emission lines in the spectral profiles of three stars in the Cygnus constellation. These stars are very hot and have masses ranging from 5 to 50 $M_{\odot}$ (solar mass). They produce high velocity and optically thick stellar winds which hide the spectral lines of the star itself. This stellar wind is not constant over time and randomly displays phases of varying activity levels, undoubtedly ejecta from the star, which no longer consumes hydrogen, but heavier elements such as helium or carbon.

WR 11 has the advantage of being the brightest Wolf-Rayet star, but it is only visible from the southern hemisphere. Here are the characteristics of this star:

WR 11 (HIP 39953), magnitude $V = 1.83$, spectral type: WC8 + O7.5III-V

RA: 08h 09' 31.95" (J2000).
DEC: -47° 20' 11.71" (J2000).

This spectrum shows broad emission lines of helium and carbon. In this example, we will use the carbon CIII line located at 5 696 Å, on the one hand because it is fairly isolated from the others and also because it is close to the maximum quantum efficiency of the camera we used (a CCD ATIK 460EX with a peak sensitivity between 520 and 550 nm).

For this type of measurement, where we want to detect small variations from one spectrum to another, with medium to high resolution spectrograph such as the Star'Ex HR, LHIRES III, eShel spectrograph, or BACHES must be used. With a Star'Ex HR or LHIRES III, perfect centering of the CIII line on the spectrograph sensor is imperative because this emission line is very broad.

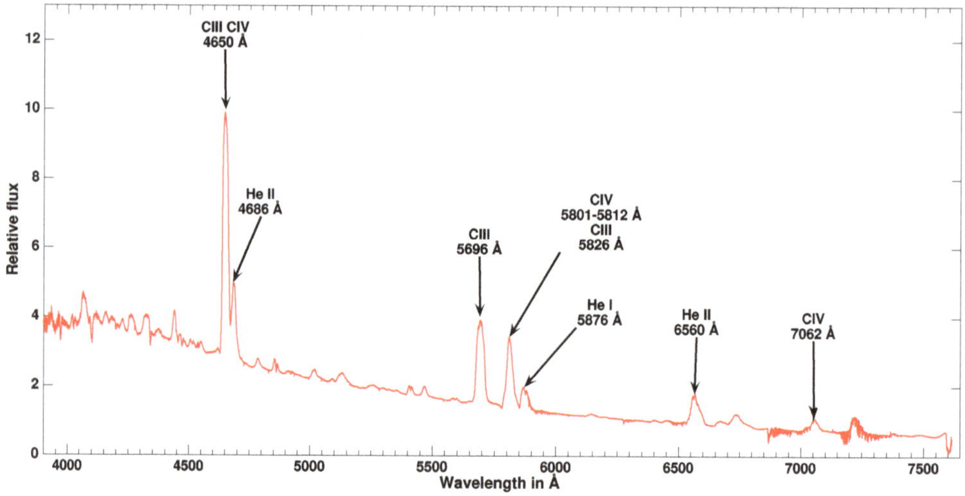

Fig. 9.4.1 — Spectrum of WR 11. Average spectrum of 396 exposures of 40 seconds. Produced with an eShel spectrograph ($R = 11\ 000$) on a 305 mm diameter Ritchey-Chrétien telescope from Chile by Lionel Mulato from the 2SPOT team on January 29, 2024. On the x-axis, the wavelength in angstroms. On the y-axis, the relative intensity compared to the continuum normalized to 1.

As with the previous examples of temporal spectral series, the best ratio between unit exposure time and signal-to-noise ratio will be determined. The series of spectra is performed over the longest possible period of time in order to increase the chances of seeing small random changes in the spectral profile.

In the case of WR 11, its magnitude allows for short exposure times, around 40 seconds. Here, WR 11 was monitored for over 5 hours on the same night, producing 396 spectra of 40 seconds each.

During processing, to improve the signal-to-noise ratio, we group the unit spectra into sets of 10 which provides a temporal resolution of 400 seconds.

For targets with shorter time evolutions, approximately about few minutes, each batch can be reduced to a total duration of around 2 to 3 minutes, but with the consequence of a degraded signal-to-noise ratio. This is necessary, for example, for CH Cyg, where variations can be on the order of minutes.

In the case of WR 11, we grouped the unit spectra into sets of 10, resulting in 39 sets.

To highlight these slight fluctuations, we will first calculate an average spectrum using all the unit spectra. This average spectrum will be more accurate and less noisy as number of spectra increases. Figure 9.4.1 shows this average profile.

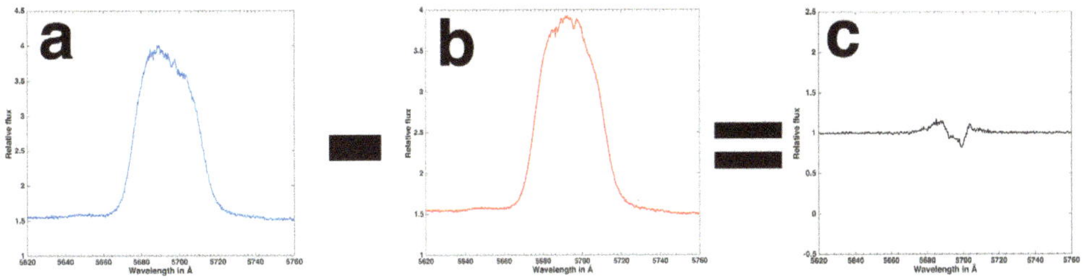

Fig. 9.4.2 — Method for calculating the residue. The spectrum (a) is a spectrum derived from the 39 batches in the series, from which the average profile (b) is subtracted, which gives the spectrum (c) resulting from the residues of this calculation.

The average spectrum profile was created using the 396 unit spectra, approximately 5 hours. This average spectrum is subtracted from each batch of spectra to obtain 39 residual spectra. This calculation can be performed using a Python script that will process all series at once. This subtraction is also possible with ISIS or Visual Spec, here is how to proceed with the ISIS software:

Go to the tab "**Tools**", then the tab "**Spectra 1**"

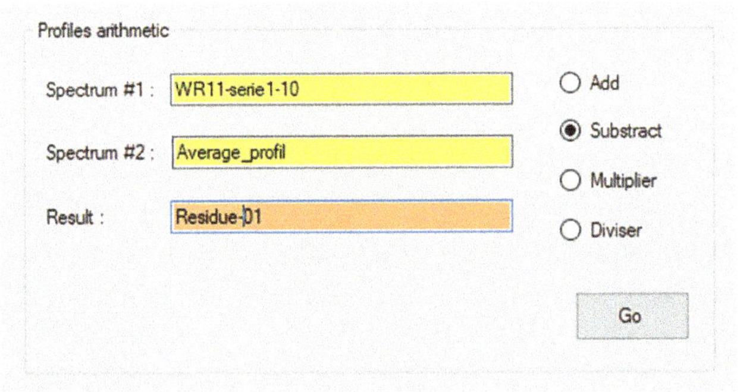

Fig. 9.4.3 — Window "Profiles arithmetic". From the ISIS software.

In the table on the right (Figure 9.4.3), enter the names of the required files:

- Spectrum #1: the name of the set of unit spectra.
- Spectrum #2: the average profile of the entire series of spectra.
- Result: the name of the file that will be the result of the subtraction.

Then click on the checkbox "**Substract**" and then on "**Go**". We repeat this as many times as we have batches. In our case, we obtained 39 residue spectra so 39 batches.

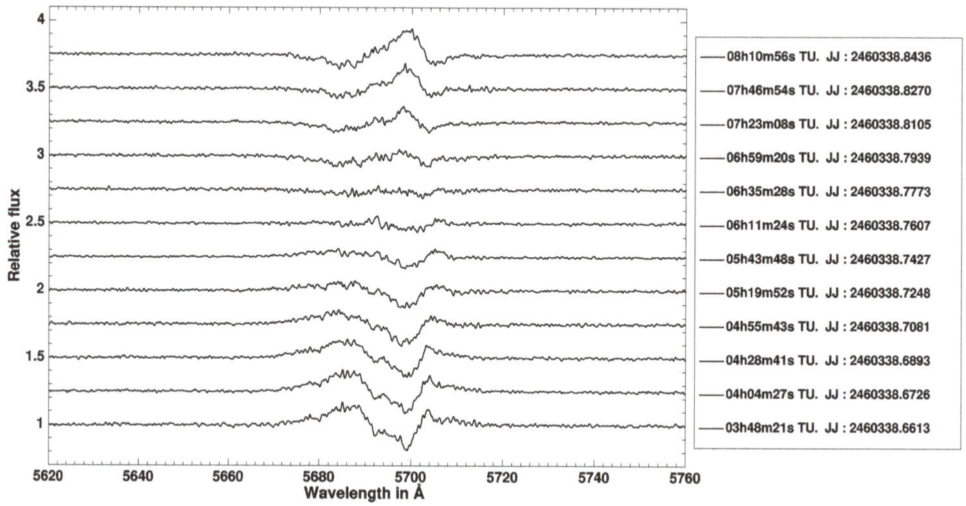

Fig. 9.4.4 — Time series on the CIII line of WR11. Presenting 12 of the 39 residue spectra. On the x-axis, the wavelength in angstroms. On the y-axis, relative intensity in relation to the continuum normalized to 1.

We can calibrate the graph in Figure 9.4.4 in velocity relative to the CIII 5 696 Å line and see the evolution in the various absorption and emission profiles.

According to a publication by W. Schmutz *et al.* (1997): "The orbital motion of gam02 Vel", WR 11 is believed to be a binary system consisting of two stars, one of type O and the other of type Wolf-Rayet whose orbital period is $P = 78.53$ days and inclined at 65 degrees to the observer's line of sight. These two stars produce a strong stellar winds which collide with each other. Depending on the phase of the period, these shocks can be viewed from various angles and different distances between the two stars.

The phase of this series of observations is determined based on the Julian day of the periastron at $JJ = 2\,450\,120.5$ (moment when the two celestial bodies are closest to each other) by the relationship:

$$\phi = \frac{(JJ_{obs} - JJ_{periast})}{P} - \left\lfloor \frac{(JJ_{obs} - JJ_{periast})}{P} \right\rfloor$$

With:

ϕ: the phase of the period.
JJ_{obs}: the Julian day of the observed spectrum.
$JJ_{periast}$: the Julian day of the last known periast, JJ = 2 450 120.5 days for the case of WR 11.
P: the orbital period in days, P = 78.53 days for WR 11.

$\lfloor \frac{(JJ_{obs} - JJ_{periast})}{P} \rfloor$: the integer part of the result.

The digital application in the case of WR 11 with the spectra obtained on January 29, 2024 gives the phase ϕ:

$$\phi = \frac{(2\,460\,338.658\,6 - 2\,450\,120.5)}{78.53} - \lfloor \frac{(2\,460\,338.658\,6 - 2\,450\,120.5)}{78.53} \rfloor = 130.1178 - 130 = 0.1178$$

Rounded to ϕ = 0.12

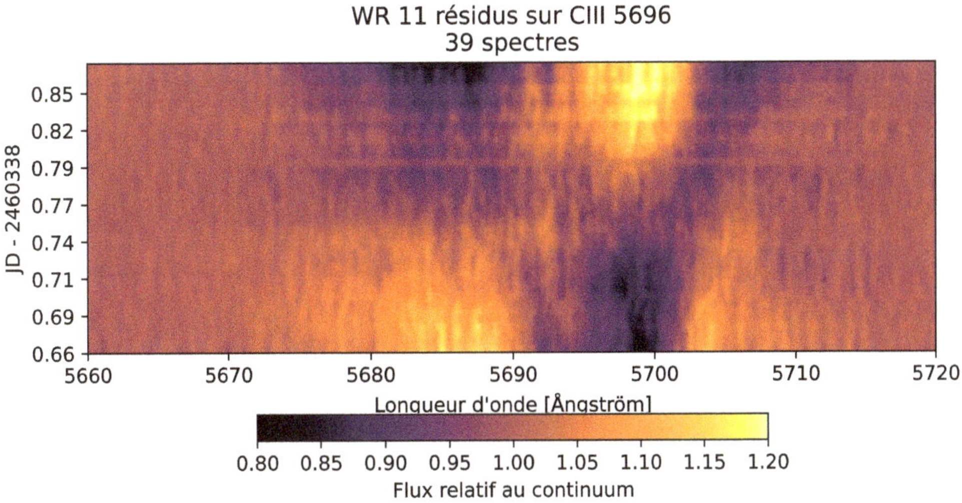

Fig. 9.4.5 — Dynamic 2D graph of residues from the CIII 5696 Å line. Performed with 39 spectra at the phase ϕ = 0.12. On the x-axis, the wavelength in angstroms. On the y-axis, the fraction of the Julian date minus 2 460 338.

The graph in Figure 9.4.5 shows developments in areas of the atmosphere of the two stars; they are not homogeneous and vary over time. Here, the duration analyzed is 5 hours and 22 minutes, on January 29, 2024, at the time of the phase ϕ = 0.12.

Tracking this type of target over several nights allows measurements to be taken on various phases of the system and to see how matter densities evolve.

Measurements of other stars are possible, such as CH Cyg or Deneb, and many more. By searching a little in the bibliography, you can find interesting targets for temporal monitoring.

9.5 Bibliography

- A new determination of the period of the beta CMa star gam peg by J.P. Sareyan, J. C. Valtier and J.M. Le Contel, (Astronomy & Astrophysics 44, 215-218 (1975)).

- The observational status of the beta cephei stars by Janet Rountree Lesh and Morris L. Aizenman (Annual Reviews Inc. 1978. 16: 215-40)

- New opacities and the origin of the beta Cephei pulsation by P. Moskalik and W.A. Dziembowski (Astronomy and Astrophysics. 256, L5-L8 (1992))

- An extensive spectroscopic time-series of three Wolf-Rayet stars. I. The lifetime of large-scale structures in the wind of WR 134 by E. J. Aldoretta *et al.* (Monthly Notices of the Royal Astronomical Society, Volume 460, Issue 3, 11 August 2016).

- The orbital motion of gam 02 Velorum by W. Schmutz *et al.* (Astronomy and Astrophysics. 328, 219-228 (1997)).

- An observational overview of pulsations in beta Cep stars and slowly pulsating B stars by P. De Cat (Astronomical Society of the Pacific, 2002ASPC**259**196D)

- Asteroseismology of the beta Cep star HD 129929 (Observations, oscillation frequencies and stellar parameters) by C. Aerts *et al.* (Astronomy and Astrophysics 415, 241–249 (2004)). DOI: 10.1051/0004-6361:20034142

- Period variation in BW Vulpeculae redux (Research Note) by A. P. Odell (Astronomy and Astrophysics 544, A28 (2012)), DOI: 10.1051/0004-6361/201219418.

- Etude à haute résolution temporelle de l'étoile pulsante BW Vulpeculae by D. Garnier, N. Nardetto, P. Mathias and D. Gillet, La lettre de l'OHP N° 20, décembre 2001.

- Nonlinear pulsational behaviour of BW Vulpeculae by P. Moskalik and J.R. Buchler. (1994) Cambridge University Press, doi/10.1017/S007418090021437X.

- Hydrodynamic models for beta Cephei variables, BW Vulpeculae revisited by A. Fokin, Ph. Mathias, E. Chapellier, D. Gillet, and N. Nardetto. (Astronomy and Astrophysics 426, 687-693 (2004)). DOI : 10.1051/0004-6361:20040418.

- Wind inhomogeneities in Wolf-Rayet stars, Unusual emission line profile variations in gam 02 Velorum by S. Lepine, T. Eversberg and A. F. J. Moffat. (The Astronomical Journal. 117:1441-1453, 1999 March).

- The Beta Cephei Stars and Their Relatives. AAVSO: https://www.aavso.org/vsots_betacep

10. Spectroscopic binaries

Spectroscopic binaries (abbreviated SB or BS) are very interesting stars to observe in spectroscopy particularly those with short periods because they are "dynamic": their spectra evolve rapidly.

Very numerous, they represent about 30 % of all visible stars in our galaxy. Some are very bright and easily accessible at medium or high spectral resolution.

The Star'Ex HR and LHIRES III spectrographs can be used (with a grating of 2 400 lines/mm) as well as Echelle-type spectrographs (eShel and BACHES) to obtain radial velocity measurements with an accuracy of around 500 $m.s^{-1}$ or less with a stable spectrograph. Some spectroscopic binaries which exhibit significant velocity amplitudes (between 50 and 200 $km.s^{-1}$) are very easy to measure. Other weaker ones of around of 1 $km.s^{-1}$ or even less, will require greater precision in terms of the quality of the spectra produced (wavelength calibration and signal-to-noise ratio).

10.1 Principle of measuring spectroscopic binaries

Spectroscopic binaries are pairs of stars orbiting around a common center of mass. They are called "spectroscopic" because their binary nature is only revealed by spectroscopy. The components of the system are either too close together to resolve them optically or one component is too faint to be visible.

Spectra produced over time will show wavelength shifts due to the Doppler-Fizeau effect. These variations in radial velocity will enable us to determine the rotation period of the binary star system, or even discover other companions with different periods of revolution.

A large number of spectra are needed covering several orbits of the stars to produce regular and consistent measurements over the entire orbital period without any "gaps".

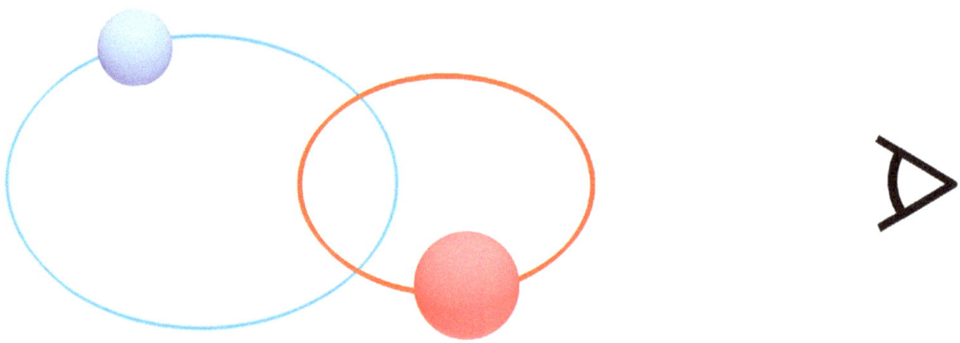

Fig. 10.1.1 — A spectroscopic binary system. The two stars (red and blue) orbit around their common center of mass. Depending on the line of sight, the two stars will move closer to or further away from the observer, producing spectra shifted in wavelength.

As with galaxies, we might be observing along a line of sight perpendicular to the plane of rotation; in these cases, we won't be able to measure anything! If the line of sight is oblique, then the inclination must be taken into account: $V\sin(i)$.

Spectroscopic binaries are classified into several categories according to the number of observable spectral lines of the same element:

• **SB1**: (BS1 in French) The spectra show a spectral shift over time in only one of the two components of the system (generally the brightest).

We encounter the same type of variations when detecting exoplanets around stars but with much lower radial velocities (< 80 m.s^{-1}). For example, the Earth's rotation around the Sun induces a radial velocity of the Sun of the order of 0.09 m.s^{-1}.

This low velocity is not be detectable with amateur equipment and at the time of writing this chapter, only the Espresso spectrograph installed on the VLT in Chile or the future HIRES, which will be mounted on the E-ELT telescope would be able to detect these exo-Earths.

• **SB2**: (BS2 in French) Spectra show in the same 2D image, the shift of the two components of the system over time: The lines are split according to the phase. This is the case, for example, with the star Mizar A, which will be studied later in this chapter.

• **SB3**: (BS3 in French) Much rarer, these spectra indicate the presence of a third companion that is not necessarily visible but which influences the final shape of the curve, discussed later in this chapter.

There are catalogs listing spectroscopic binary stars: for example the SB 9, published in 2004: "**SB9: The ninth catalogue of spectroscopic binary orbits**" by D. Pourbaix *et al.* (A&A 424, 727–732 (2004))

10.2 A bit of history

The star pair Alcor and Mizar has been known since ancient times and it was around 1650 that the astronomer Giovanni Battista Riccioli (1598-1671) discovered that Mizar itself is a double star with a companion located 14.4 arc seconds away forming the pair Mizar A and Mizar B.

The first spectroscopic binary star was observed by the German astronomer Hermann Carl Vogel (1841-1907) in November 1889 at the Potsdam Observatory. He was able to measure variations in velocity of Algol and estimate the diameter of this star and its main companion (it is actually a much more complex system with at least three components). Almost simultaneously in the same year, Edward Charles Pickering (1846-1919) indicates that one of his colleagues at Harvard University, Antonia C. Maury (1866-1952) discovered that the pair Mizar A and Mizar B, although visually binary, also contains a spectroscopic binary. The study of Mizar A's spectrum showed a splitting of the calcium K line over time.

By 1955, 1 350 spectroscopic binaries had been identified and this number has only grown, now reaching more than 180 000 (source SIMBAD). Statistical studies conducted by Carlos Jaschek and Ana E. Gomez in 1970 (PASP, 82, 809) indicate that nearly 50 % of main sequence stars are binary.

The accuracy and resolution of spectrographs improved significantly in the 1990s, enabling the detection of companions much smaller than a star. This is how, in 1995, Michel Mayor and Didier Queloz discovered the first exoplanet orbiting the star 51 Peg (magnitude V = 5.46) with a period of 4.23 days. The observation was made using the 1.93 m telescope at the Haute-Provence Observatory (OHP) in France using the Elodie spectrograph: resolution R = 45 000. The amplitude of radial velocity variations from exoplanets is very small. In the case of 51 Peg, this difference is of the order of $\pm$ 50 m.s^{-1}.

The two astrophysicists at OHP were rewarded for this discovery by receiving the Nobel Prize in Physics in 2019.

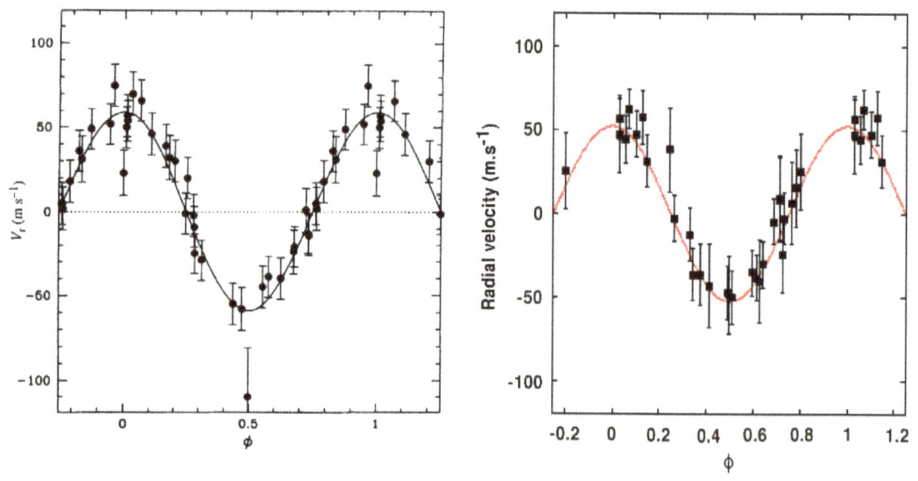

Fig. 10.2.1 — Period of 51 Peg in radial velocity. On the left, the historical graph from Michel Mayor and Didier Queloz's 1995 publication (excerpt from the article "A Jupiter-mass companion to a solar-type star" published in the journal *Nature*). On the right, measurements taken by Christian Buil (amateur) using a 28 cm telescope and a VHIRES spectrograph (R = 50 000). On the x-axis, the period in phase, on the y-axis, the amplitude of the velocity in m.s^{-1}.

In 2011, French amateur astronomer Christian Buil, a pioneer in amateur spectroscopy, re-measured this first exoplanet using a 28 cm diameter Schmidt-Cassegrain telescope and a bench mounted fiber optic spectrograph produced for the occasion, the VHIRES (R = 50 000). Christian produced this series of measurements at several observation sites: mostly on his balcony in the suburbs of Toulouse in France, but also from the T60 at Pic du Midi and at the Haute-Provence Observatory during a spectroscopy internship.

To measure such low velocities, we use a function known as cross-correlation (CCF for *cross correlation function*) by comparing several thousand spectral lines from a known and stable star with another of the same spectral type to determine the wavelength shift.

It is the use of a large number of spectral lines that allows for this precision based on statistical calculation. This function therefore applies more to stars of type F, G, K, or even M. The principle will not be described in this book because it would take almost an entire book to explain the procedure to be continued for acquisitions, treatment, prerequisites, making a mask, etc. But please note that this feature exists in the ISIS software (tab "**Misc**", then "**radial velocity**"). Many Python libraries also offer this type of calculation.

10.3 Observe a spectroscopic binary star

A very important point when starting to perform spectroscopic binaries measurements, is to have a replicable protocol for observing and reducing spectra. Always use the same spectrograph on the same optics with the same camera; a fixed installation is of course preferable. Always take a calibration image immediately after the last exposure and without moving the telescope.

During the treatment phase, carefully check the accuracy of your spectral calibration using, for example, the atmospheric lines surrounding the $H\alpha$ line to verify any discrepancy between their theoretical positions and your observation. A measurement dispersion of around 0.01 Å is required across your entire spectrum series and do not hesitate to repeat the treatment if you notice significant discrepancies.

Of course, the accuracy needed will depend on the values of the target's radial velocity variations. You will not need the same level of accuracy if you plan to measure an amplitude of ±150 km.s^{-1} compared with ±10 km.s^{-1}.

By setting up equipment that never changes, combined with a consistent acquisition and processing protocol, it is far more probable that an observer will create a relaible set of measurements.

Start with luminous targets and be patient because obtaining a radial velocity curve takes time. It typically takes several nights of observation, sometimes several weeks, months, or years... But nothing prevents you from processing your spectra as you go along and thus observe the evolution of the spectral profile over time. Obtaining a complete curve and measuring a period requires a large number of measurement points.

A selection of targets to observe can be easily obtained using SIMBAD:

https://simbad.u-strasbg.fr/simbad/

in the tab "**Criteria query**", enter, for example:

maintype = "SB*" & Vmag<4

Nom	AD (J2000)	DEC (J2000)	Vr	Mag V	Type spectral
alf Cma	06h 45' 08"	-16° 42' 58"	-5.50	1.46	A0mA1Va
zet Ori A	05h 40' 45"	-01° 56' 33"	18.10	1.90	O9.2IbvarNwk
alf Gem A	07h 34' 35"	+31° 53' 17"	6.00	1.90	A1.5IV+
bet Aur	05h 59' 31"	+44° 56' 50"	-15.75	1.90	A1IV-Vp
gam Gem	06h 37' 42"	+16° 23' 57"	-12.63	1.92	A1.5IV+
alf Pav	20h 25' 38"	-56° 44' 06"	2.00	1.92	B2IV
bet Per	03h 08' 10"	+40° 57' 20"	4.00	2.12	B8V
zet01 Uma	13h 23' 55"	+54° 55' 31"	-5.60	2.22	A1.5Vas
alf CrB	15h 34' 41"	+26° 42' 52"	1.70	2.24	A1IV
del Sco	16h 00' 20"	-22° 37' 18"	-6.00	2.32	B0.3IV
kap Vel	09h 22' 06"	-55° 00' 38"	21.90	2.47	B2IV
zet Cen	13h 55' 32"	-47° 17' 18"	6.50	2.55	B2.5IV
tet Car	10h 42' 57"	-64° 23' 40"	459.70	2.76	B0Vp
iot Ori	05h 35' 25"	-05° 54' 35"	27.60	2.77	O9IIIvar
pi. Sco	15h 58' 51"	-26° 06' 50"	-7.40	2.91	B1V+B2:V:
eps Aur	05 01' 58"	+43° 49' 23"	-10.40	2.99	A9Ia
bet Tri	02h 09' 32"	+34° 59' 14"	12.30	3.00	A5III
iot UMa A	08h 59' 12"	+48° 02' 30"	9.00	3.13	A7V(n)
del Her	17h 15' 01"	+24° 50' 21"	-40.0	3.13	A1IVn
tet Aql	20h 11' 18"	-00° 49' 17"	-28.02	3.22	B9.5III
a Car	09h 10' 58"	-58° 58" 00'	23.30	3.40	B2IV-V
lam Tau	04h 00' 40"	+12° 29' 25"	17.80	3.41	B4IV
bet Lyr	18h 50' 04"	+33° 21' 45"	2.20	3.42	B8.5Ib-II
eps Gru	22h 48' 33"	-51° 19' 00"	-0.40	3.47	A2IVn
ksi Ser	17h 37' 35"	-15° 23' 54"	-45.85	3.52	A9IIIpSr
ups04 Eri	04h 17' 53"	-33° 47' 54"	17.60	3.56	B8V+B9.5V?
eta Ori A	05h 24' 28"	-02° 23' 49"	?	3.59	B1V
27 Tau	03h 49' 09"	+24° 03' 12"	8.50	3.63	B8III
tau Lib	15h 38' 39"	-29° 46' 39"	3.20	3.64	B2.5V
alf Dra	14h 04' 23"	+64° 22' 33"	-13.0	3.68	A0III
pi.04 Ori	04h 51' 12"	+05° 36' 18"	23.3	3.68	B2III
pi.05 Ori	04h 54' 15"	+02° 26' 26"	24.20	3.73	B2III
p Vel	10h 37' 18"	-48° 13' 32"	21.20	3.84	A3/5
rho Sco	15h 56' 53"	-29° 12' 50"	-0.40	3.86	B2IV-V
gam Mus	12h 32' 28"	-72° 07' 58"	7.78	3.88	B5V+A:
eta Vir	12h 19' 54"	-00° 40' 00"	5.24	3.90	A2IV
omi Per	03h 44' 19"	+32° 17' 17"	12.20	3.91	B1III
eps Her	17h 00' 17"	+30° 55' 35"	-25.10	3.92	A0V
chi Lup	15h 50' 57"	-33° 37' 37"	-16.30	3.95	B9III

Fig. 10.3.1 — Brightest spectroscopic binary stars. Extract from the SIMBAD database. The total number of stars for this query is around 90. The Vr column indicates the star system's mean velocity relative to an observer on Earth.

This query asks SIMBAD to display spectroscopic binaries stars (abbreviation "SB*") whose magnitude is less than 10. The use of this tool has been described in section 2.2.

SIMBAD's response will indicate that there are more than 16 000 stars of this type. If we restrict the results to stars with magnitude $V < 4$ and only spectral types O, A, and B in order to obtain spectra without too many spectral lines (except those of hydrogen), we obtain a list of around thirty stars.

This list is already a good starting point for beginning to study spectroscopic binaries. Some of them will be described in the following pages.

10.4 Practical case study based on Mizar A

Mizar A (zet01 UMa) is an easy target to reach because of its brightness (magnitude $V = 2.22$) which means shorter exposures with a high-resolution spectrograph. In addition, it is visible almost all year round from the northern hemisphere. This is an SB2 type binary star, that is, its spectrum shows both components of the system. Its period is $P = 20.538\,35$ days. It will therefore be necessary to track this star by taking spectra over a period of one month or more, at a rate of one or two spectra per night.

At least fifteen observations will be necessary in order to reconstruct the entire period of this binary.

In this example, we used 14 spectra that were taken over a period of 478 days. Initially, we did not specifically intend to follow this star regularly and depending on the night, we observed this target completely at random, which is why it took so long to obtain this series. But you can easily spend much less time observing to obtain the same number of spectra.

Determining the exposure time is important in order to have a good signal-to-noise ratio (number of ADUs per unit exposure between 30 000 and 40 000).

Each spectrum will be processed individually, paying particular attention to the accuracy of wavelength calibration. Then, for each one, it will be necessary to:

- Remove atmospheric lines if spectra are taken on the $H\alpha$ line. The removal of these lines was explained in section 5.6.

If the observation has been made on other lines, there is no need to remove them as there are not necessarily any! (such as $H\beta$).

• Correct each spectrum to the barycentric velocity as we saw in section 3.7: ± 30 km.s^{-1} ...
Failure to do this would result in a huge error in relation to what we are trying to measure!

• Flatten (intensity = 1) the continuum of the star near the line to be measured as described in section 5.6.

• Measure the wavelength position of the spectral line by modelling the spectral profile with a Gaussian function. This procedure is described in section 3.8.

• Measuring the Doppler-Fizeau shift $\Delta\lambda$ of each spectrum relative to the value of the wavelength at rest of the line (λ_0) by applying the simplified equation in section 3.4 and deduce its radial velocity using the relationship: $V_r = cz = c\frac{\Delta\lambda}{\lambda_0}$.

The sign of $\Delta\lambda$ is important.

• Each measurement point is arranged in a spreadsheet with two columns:

Column n° 1: the Julian day of the observation.
Column n° 2: the measured value of the radial velocity (in km.s^{-1}).

10.5 Order the spectra according to phase

The series of measurements you took of Mizar, certainly over several weeks or months, will cover several periods and must be aggregated into a single phase whose value varies between 0 and 1.

0: the beginning of the period P.
1: the end of the period P (in the case of Mizar $P = 20.538\,35$ days).

The phase is denoted by the Greek letter Φ (Phi). The phase $\Phi = 0.5$ corresponds to half of the period $P = 10.269$ days in the case of Mizar A. The equation used to classify a series of spectra by phase is given in section 9.4.

This spectrum reclassification work can be performed automatically by many software programs but also with a Python script. Of course, you need to know the period of the binary system being studied. Here is how to proceed with ISIS:

• Create a text file with two columns containing the name of the spectrum and the Julian day date corresponding to that spectrum.

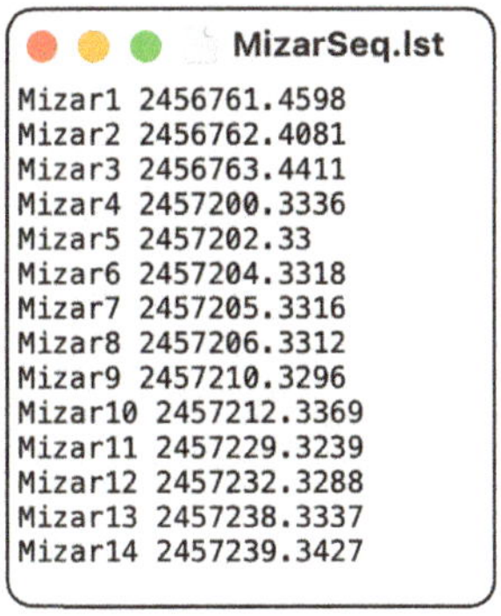
MizarSeq.lst

```
Mizar1 2456761.4598
Mizar2 2456762.4081
Mizar3 2456763.4411
Mizar4 2457200.3336
Mizar5 2457202.33
Mizar6 2457204.3318
Mizar7 2457205.3316
Mizar8 2457206.3312
Mizar9 2457210.3296
Mizar10 2457212.3369
Mizar11 2457229.3239
Mizar12 2457232.3288
Mizar13 2457238.3337
Mizar14 2457239.3427
```

Fig. 10.5.1 — Contents of the .lst text file. Including the 2 columns with the name of the spectrum and the Julian day.

This file lists all the spectra produced in chronological order (acquisition date). Be sure to use a full stop "**.**" instead of a comma to separate the decimal part from the Julian day. It is saved with the .lst extension in the working directory specified in the ISIS software preferences (see Figure 10.5.1).

• Select the tab "**Misc**" then "**Animation**". On the right side of this page, you will find a window "**Conversion to phase file**".

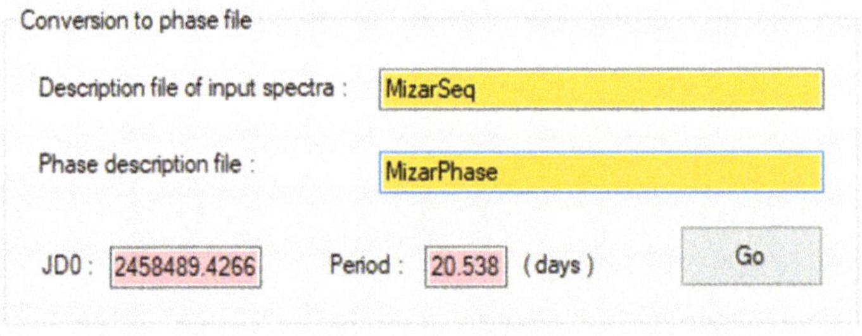

Fig. 10.5.2 — Window "Conversion to phase file". From the ISIS software.

Fill in the following fields:

• Description file of input spectra: the name of the file created in the previous step containing the list of spectra sorted in chronological order.

• Phase description file: the name of the file in which the spectra will be classified by phase after calculation by ISIS.

• JDO: the Julian day of the first spectrum in the sequence. This may also be the day of the periastron of the system if it is known or even the photometric maximum of a period.

• Period: the value of the period P in days (in the case of Mizar A, the value is $P = 20.53835$ days).

Click on "**GO**", The phase text file is generated by ISIS.

Vocabulary note: The words periastron and apastron refer to the closest and furthest points from a celestial body. Here we will make a slight linguistic deviation by replacing the star with the center of gravity of the binary system.

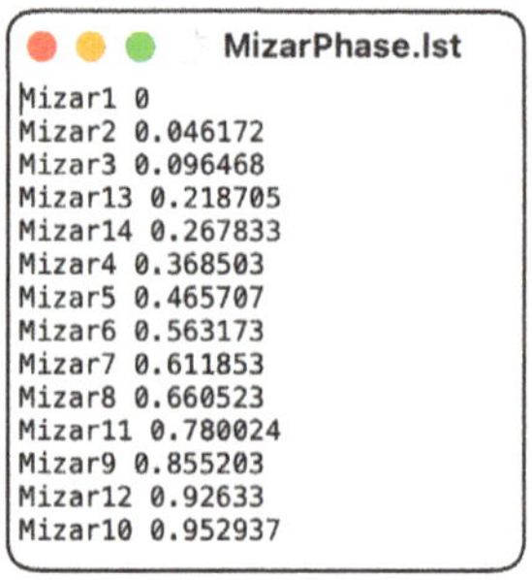
MizarPhase.lst

```
Mizar1 0
Mizar2 0.046172
Mizar3 0.096468
Mizar13 0.218705
Mizar14 0.267833
Mizar4 0.368503
Mizar5 0.465707
Mizar6 0.563173
Mizar7 0.611853
Mizar8 0.660523
Mizar11 0.780024
Mizar9 0.855203
Mizar12 0.92633
Mizar10 0.952937
```

Fig. 10.5.3 — Result of the phase calculation in the .lst file. Generated by ISIS.

This file contains two columns sorted chronologically: the name of the spectrum followed by the phase value (from 0 to 1).

This example shows that the Mizar13 and Mizar14 spectra are interspersed in phase between Mizar3 and Mizar4 (see Figure 10.5.3).

10.6 The spectral evolution of Mizar A

Now that your spectra are sorted by phase rather than chronological order of acquisition, you can view the evolution of the spectral profile over time for a single period.

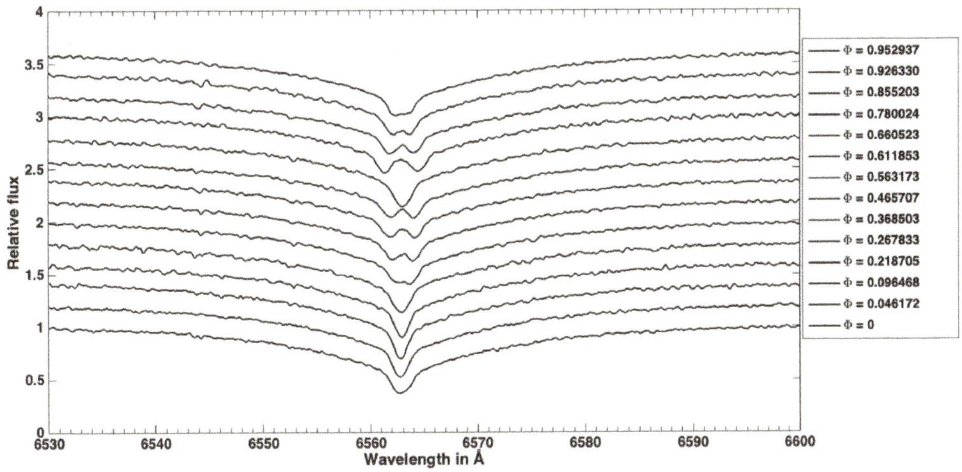

Fig. 10.6.1 — Spectral series for Mizar A. The 14 spectra centered on the $H\alpha$ line are classified by phase (0 to 1). Produced with an eShel spectrograph (R = 11 000) on a 406 mm diameter Ritchey-Chretien telescope and an ATIK 460EX camera. On the x-axis, wavelength in Angströms. On the y-axis, relative intensity to the continuum normalized to 1.

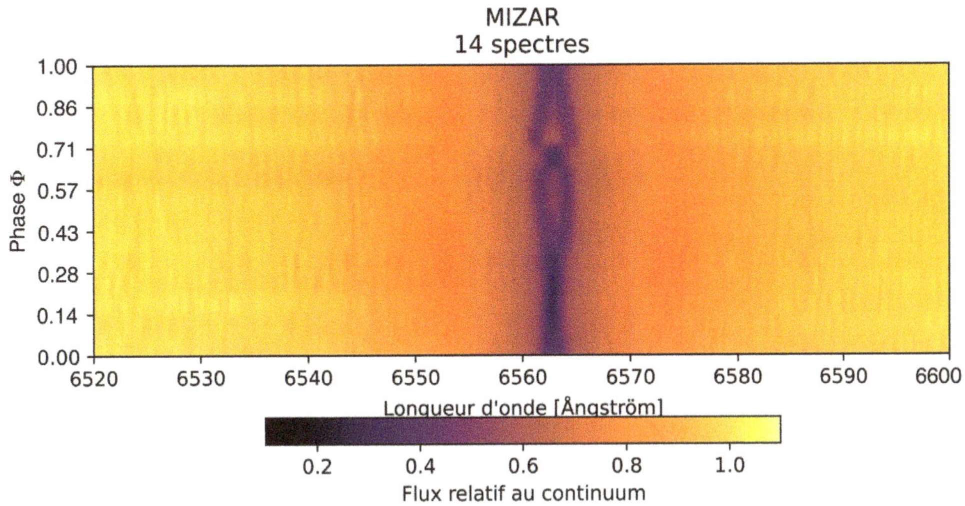

Fig. 10.6.2 — Dynamic 2D graph showing the evolution of Mizar A. Created using the Python script described in section 9.1 using the 14 spectra created previously. On the x-axis, wavelength in Angströms. On the y-axis, the phase ϕ of the period. The intensity of the spectrum is represented according to the colour scale indicated.

The evolution of the spectral profile clearly shows that Mizar A is of the SB2 type because we see the splitting of the $H\alpha$ line over time. Direct measurement of the velocities of both components is possible in this case.

A dynamic 2D graph provides a better representation of the movements of the two components evolving as if in a ballet and repeating identically period after period.

If the phenomenon is not identical from one period to another, we can assume the presence of another companion in the system and determine a secondary period to be added to the first.

10.7 Measuring the period of a spectroscopic binary: the periodogram

When we study a spectroscopic binary star, we generate a large number of spectra on which we can measure radial velocity fluctuations. The periodogram indicates the most probable periods based on the values entered.

With a series of measurements comprising 2 parameters (the date of the spectrum and the measured radial velocity), it is possible to determine one or more periods of a system.

This is done by applying the Lomb-Scargle algorithm which allows a periodogram to be created from data that are not produced on a regular basis.

Python libraries such as astropy also offer this well-documented function:

https://docs.astropy.org/en/stable/timeseries/lombscargle.html

This algorithm is also available in ISIS. Here is the method to follow, using a series of measurements on the star HIP 45080 as an example (a Car magnitude $V = 3.40$). First, we start by creating a text file with two columns:

• Column n° 1: the observation date in Julian days. You can use only a part of the date which makes the results easier to read. This is referred to as MJD (*Modified Julian Day*). Remember to mention that you intend to do this!

For example, instead of taking the entire date 2 460 289.632 2, we can just settle for 289.632 2 (2 460 289.632 2 – 2 460 000) or even 89.632 2.

• Column n° 2: the radial velocity measured relative to the rest value of the reference line. For example, if for the $H\alpha$ line (6 562.82 Å at rest) the measurement is 6 562.52 Å, we find a velocity of:

$$Vr = c\left(\frac{\lambda_{Obs} - \lambda_0}{\lambda_0}\right) = 299\,792.458\;\frac{(6\,562.52 - 6\,562.82)}{6\,562.82} = -13.70\ \text{km.s}^{-1}$$

This file is saved with the .dat extension.

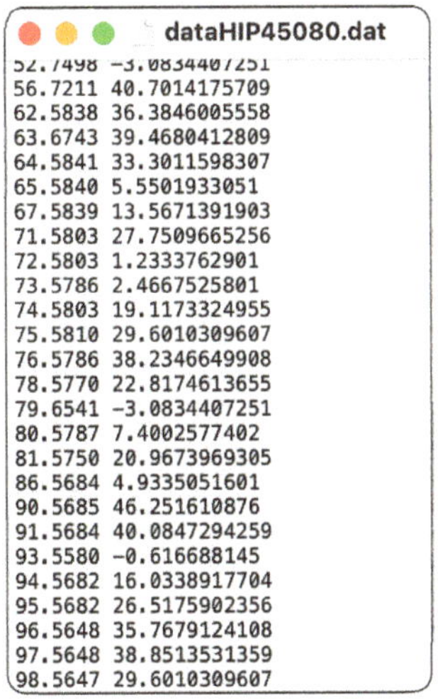

dataHIP45080.dat

52.7498	-3.0834407251
56.7211	40.7014175709
62.5838	36.3846005558
63.6743	39.4680412809
64.5841	33.3011598307
65.5840	5.5501933051
67.5839	13.5671391903
71.5803	27.7509665256
72.5803	1.2333762901
73.5786	2.4667525801
74.5803	19.1173324955
75.5810	29.6010309607
76.5786	38.2346649908
78.5770	22.8174613655
79.6541	-3.0834407251
80.5787	7.4002577402
81.5750	20.9673969305
86.5684	4.9335051601
90.5685	46.251610876
91.5684	40.0847294259
93.5580	-0.616688145
94.5682	16.0338917704
95.5682	26.5175902356
96.5648	35.7679124108
97.5648	38.8513531359
98.5647	29.6010309607

Fig. 10.7.1 — Example of measurements on HIP 45080. Extract from the text file in .dat format containing a total of 69 measurement points. Spectra automatically produced from Chile by the 2SPOT team between November 13, 2023, and April 4, 2024.

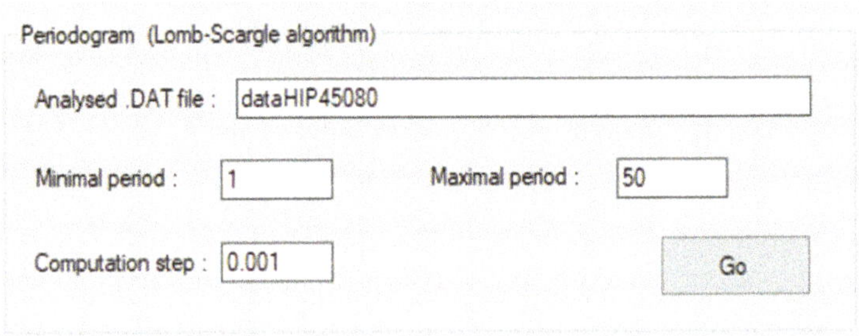

Fig. 10.7.2 — Periodogram settings window. From the ISIS software.

This file should be placed in the current ISIS working directory. Then select the tab "**Misc**" then "**radial velocity**". On the left side of the window, you will find a section "**Periodogram (Lomb-Scargle algorithm)**".

Fill in the fields in Figure 10.7.2 with:

• Analysed .DAT file: the name of the text file containing the measurements.

• Minimal period and maximal period: the range of analysis for the probable period in days.

• Computation step: indicate the minimum value of the step in days. Here, the value of 0.001 days corresponds to a step of 1.44 minutes.

Then click on "**GO**". The software displays the results just below.

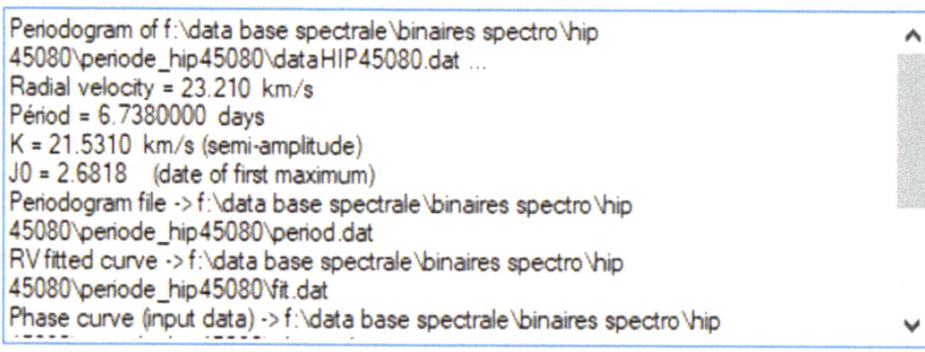

Fig. 10.7.3 — Periodogram results window. From the ISIS software.

Figure 10.7.3 indicates the results that are displayed:

• The radial velocity of the star: On average, it is moving away from us at a velocity of 23.21 km.s^{-1}. The value indicated by SIMBAD is 23.30 km.s^{-1}, that is a difference of 0.09 km.s^{-1}.

• The calculated period P = 6.738 days. The generally accepted value found in publications is 6.7447 days, that is a difference of 0.0067 day (a little less than 10 minutes).

• The semi-amplitude coefficient K: K = 21.5310 km.s^{-1}, the value in some publications is 21.50 km.s^{-1}, a difference of 0.031 km.s^{-1}.

• J_0 = 2.6818 days, date of the first maximum velocity from the series of spectra. (To get the exact date, you need to add back the 2 460 000 that we removed from the original text file).

ISIS also generates five .dat files that are useful for reconstructing various curves from the data in the initial text file. You can import these files into a spreadsheet to create the associated curves:

• phase.dat: to generate the phase curve of the raw input data: the first column contains the phase value ϕ, the second is the radial velocity in m.s^{-1}. It is necessary to sort the phase data in the first column in ascending order.

• phase2.dat: the values of the theoretical curve for the period with two columns similar to the phase.dat file (it is C that will be used for the O-C calculation).

• deltaphase.dat: the O-C error of the phase curve (O: the observed value, C: the calculated value) of the radial velocity values. The first column is the phase value, the second column shows the O-C error of the radial velocity.

• fit.dat: the adjusted velocity curve for all observations with the Julian day date in the first column and the adjusted radial velocity in the second.

• period.dat: contains the periodogram data with the period value in days in the first column and in the second, the power of the probability of a period.

It is this last file that is interesting to trace and which will show us the probable or possible periods.

The graph in Figure 10.7.4 shows a maximum probability at a period of P = 6.738 days, which is close to the true period of HIP 45080 within 0.1%. Other less intense probability peaks are also present (at 1.18, 1.42 and 3.38 days) and may correspond to the presence of another companion.

But it is probably a bias or error in the series of measurements!

A bias may occur if we perform a spectrum every night, at almost always at the same time or at regular intervals. This creates a false period that is not due to the target but to the observation made with the same time interval between two measurements.

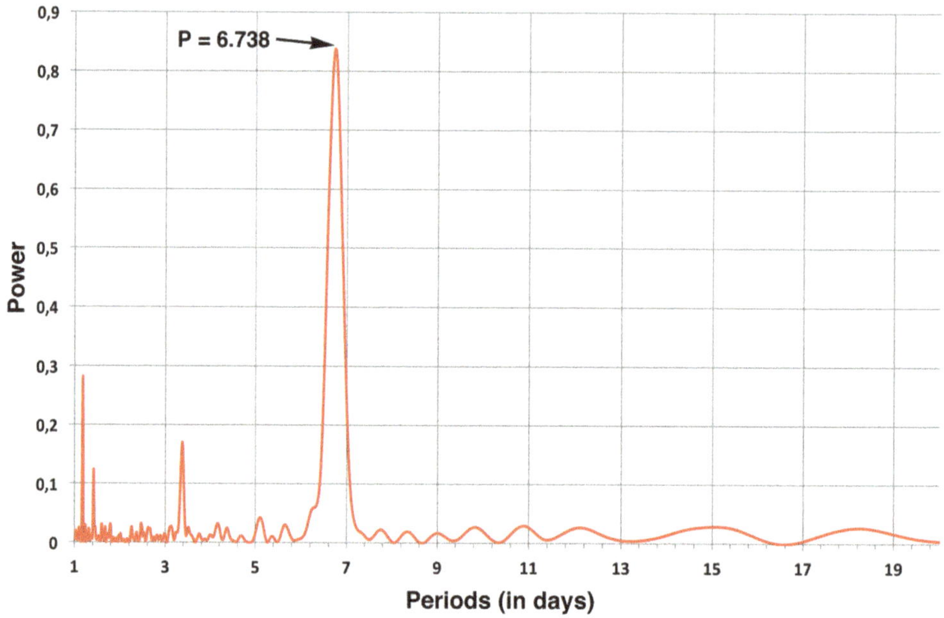

Fig. 10.7.4 — Periodogram of HIP 45080 measurements. Produced using the 69 measurement points in Figure 10.7.1. On the x-axis, the periods in days. On the y-axis, the power of probability of the existence of a period.

This period alias is determined by the following relationship:

$$P_{Alias} = \frac{1}{\left(\frac{1}{P_{obs}}\right) - \left(\frac{1}{P_{target}}\right)}$$

With:

P_{obs}: the observation period between two spectra.

P_{target}: the target period calculated by the periodogram.

P_{Alias}: the value of the alias for the period in days.

Digital application based on the periodogram in the Figure 10.7.4 with P_{target} = 6.738 days. It is estimated that one spectrum was produced every night at almost the same time, either $P_{obs} \approx 1$ day, we obtain:

$$P_{Alias} = \frac{1}{\left(\frac{1}{1}\right) - \left(\frac{1}{6.738}\right)} = 1.174 \text{ day}$$

This is a value very close to the value of 1.18 days in the Figure 10.7.4. It should therefore be excluded from the possibilities for this star's period. The other two values are also aliases because, upon closer examination of all 69 measures, we see that they often correspond to an observation gap between two spectra.

10.8 The eccentricity of an orbit

The orbits of spectroscopic binaries are rarely circular and are in elliptical paths that are more or less elongated around the system's center of gravity. The center of gravity of a binary system occupies one of the two foci of the ellipse.

It was in 1609 that Johannes Kepler (1571-1630) published the three relationships that describe the motion of the planets around the Sun, measuring the eccentricity of Mars' orbit based on Tycho-Brahe's readings (1546-1601).

The planet Mars has a fairly low eccentricity: $e = 0.093$. In the case of spectroscopic binaries, much higher values may be encountered, but by definition < 1.

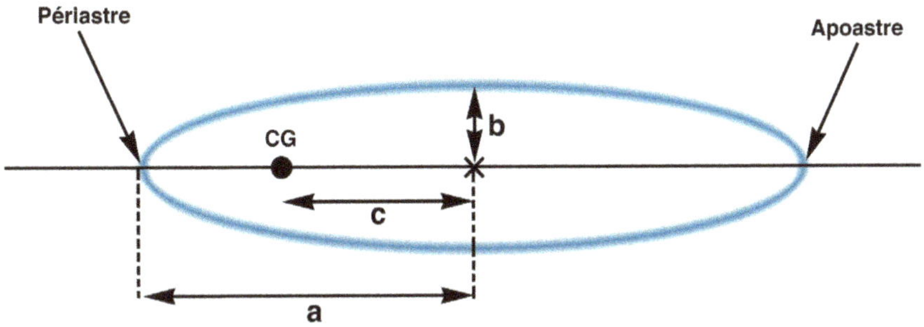

Fig. 10.8.1 — Characterization of the eccentricity of an ellipse. a = length of the semi-major axis of the ellipse, b = length of the semi-minor axis of the ellipse and c = distance between the center of the ellipse and one of the two foci.

The parameter e defines the eccentricity of the ellipse according to the relationship:

$$e = \frac{c}{a}$$

c is also defined by the relationship:

$$c = \sqrt{a^2 - b^2}$$

When $0 < e < 1$, the orbit is elliptical and periodic. These are the ranges of values that can be found on spectroscopic binaries.

When $e = 0$, $c = 0$, the orbit is perfectly circular. The two foci of the ellipse are coincident. Values close to zero are often observed in systems hosting exoplanets. For example, Jupiter's eccentricity of its revolution around the Sun is $e = 0.048$.

When $e = 1$, The trajectory is parabolic and non-periodic, and therefore not closed.

When $e > 1$, The trajectory is hyperbolic and non-periodic, therefore non-closed. This is the case with certain comets that are "just passing through".

Spectroscopic binaries orbiting around a common center of gravity therefore have an eccentricity between 0 and strictly less than 1.

We will describe two examples of binaries with very different eccentricities later on.

The eccentricity of the orbit also induces different velocities depending on the position of the companion rotating around the system's center of gravity. Kepler's second law, commonly known as the "law of areas" indicates that the area swept between a rotating object and the center of gravity of the system over time is constant regardless of the position of the object in its orbit.

This implies that the orbital velocity varies in the case of an elliptical orbit. It is at its maximum when the object passes the periastron (minimum distance to the system's center of gravity) and it is minimal at apastron (maximum distance to the system's center of gravity).

The velocity profiles therefore vary considerably depending on the value of the eccentricity e of the system.

In the case of low eccentricity, a series of regular spectral measurements will be sufficient to accurately describe the velocities variations during the period.

On the other hand, in the case of high eccentricity, it will be necessary to measure a large velocity variation over a relatively short period of time (depending, of course, on the system period). The difficulty will be obtaining enough measurement points at the moment of periastron passage.

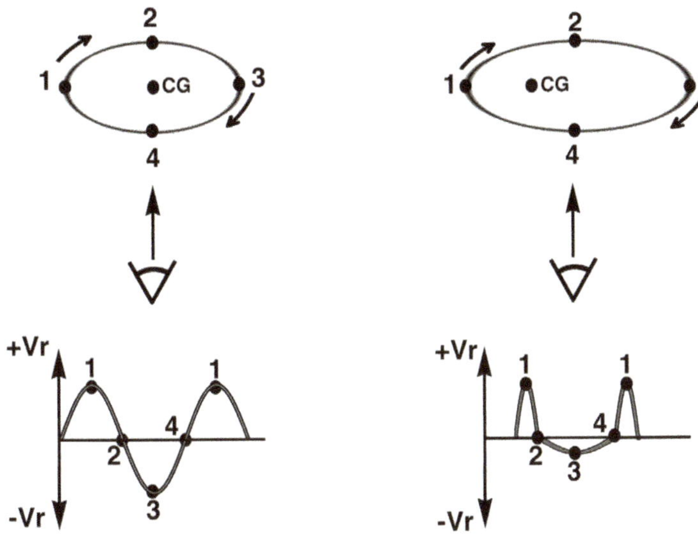

Fig. 10.8.2 — Radial velocity profile according to eccentricity *e*. On the left, a circular orbit: the rotational velocity is constant during one revolution. In spectroscopy, a sinusoidal velocity curve is observed.

On the right, an elliptical orbit with different velocities during the period of revolution due to variations in the distance between the object and the center of gravity of the system. This velocity curve is not a sine wave. (Be aware with the inclination! It is in reality $V\sin(i)$).

Note that the velocity measured here is made according to the line of sight between the star and the observer. We measure the velocity $V\sin(i)$ as shown in the Figure 8.0.1 which is necessarily smaller than the real velocity except when the orbital plane is in the observer's line of sight.

10.9 HIP 45080, a binary star with low eccentricity

Its common name is a Car (a Carina), visible from the southern hemisphere. Its magnitude is $V = 3.40$ and its spectral type B2IV-V.

We observed this star from Chile with our 2SPOT team by producing 69 spectra over 5 months with an eShel spectrograph on a 305 mm diameter Ritchey-Chretien telescope. Here are the coordinates of this spectroscopic binary:

RA: 09h 10' 58.08" (J2000)
DEC: -58° 58' 00.84" (J2000)

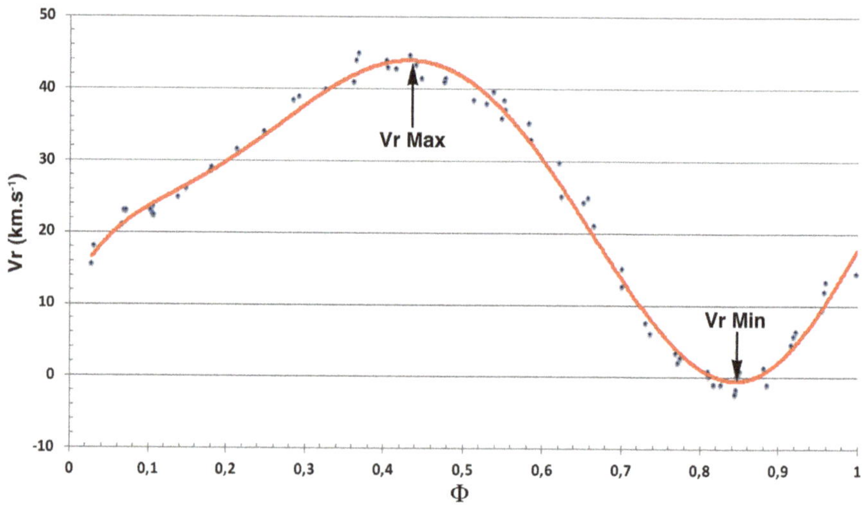

Fig. 10.9.1 — Velocity variation of HIP 45080. Produced from 69 spectra grouped on one phase Φ. On the *x*-axis, the phase Φ, on the *y*-axis, velocity in km.s^{-1}.

The data we have produced covers the entire phase in a consistent manner. The graph of velocities over a period shows an almost sinusoidal curve, signature of low eccentricity ($e = 0.16$ for HIP 45080). In this case, the semi-amplitude coefficient K can be easily found by calculating the difference between the minimum and maximum values of the radial velocity with:

$$K = \frac{(Vr_{Max} - Vr_{Min})}{2}$$

On the other hand, K is more complicated to determine with greater eccentricity because the curve is not symmetrical.

10.10 HIP 26241, a highly eccentric binary star

Also known by the common name of iot Ori (iota Orion), close to the famous M42 in Orion, it has a magnitude $V = 2.77$ and a spectral type 09III. It is therefore a very hot star with hydrogen and helium lines. The observations were made from Chile using the same equipment as in the previous example on HIP 45080 over a period of 5 months, producing 66 spectra.

Here are the coordinates of this spectroscopic binary:

RA: 05h 35’ 25.98” (J2000)
DEC: -05° 54’ 35.60” (J2000)

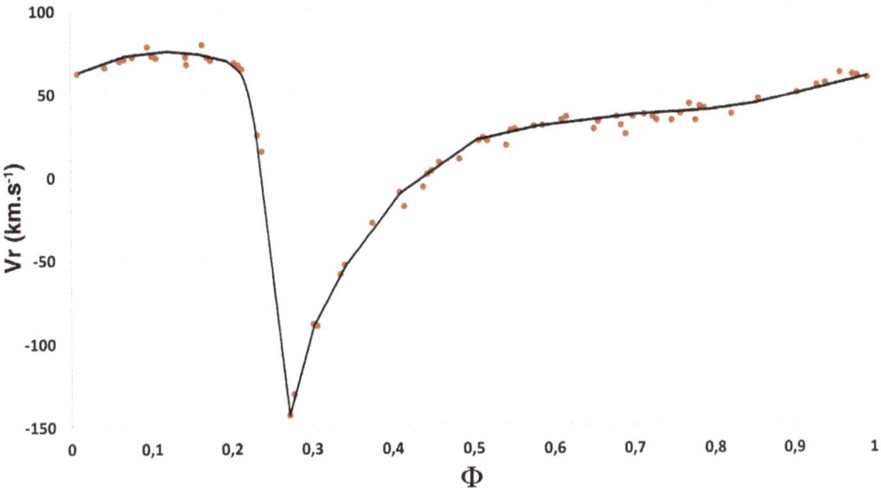

Fig. 10.10.1 — Variation in velocities of HIP 26241. Produced from 66 spectra grouped on one phase Φ.

The velocity graph is quite different from the previous example. HIP 26241 has an eccentricity $e = 0.76$. The curve no longer has a sinusoidal shape at all. The velocity at periastron can be measured at around 145 $km.s^{-1}$ (maximum velocity at phase $\Phi = 0.27$). The velocity at the periastron is more difficult to measure with certainty.

The data we have produced shows that there are many missing spectra between phase $\Phi = 0.2$ and $\Phi = 0.4$, phases during which speed variations are very significant over a relatively short period of time. The maximum velocity value we measured is unreliable due to a lack of other measurements at those specific moments in the phase. The period of this star is $P = 29.1338$ days. The radial velocity of its proper motion of 27.60 $km.s^{-1}$ (SIMBAD data).

By plotting velocity curves of spectroscopic binaries stars, you will encounter as many special cases as there are objects.

This is due to the diversity of the orbits of these stars as seen from the observer's angle of view ($V\sin(i)$) and the complexity of each system.

10.11 3D printing of the period of a spectroscopic binary

To conclude this chapter, we will discover how to print a 3D representation of the period of a spectroscopic binary star.

It is more the playful than the scientific aspect that is addressed here, in order to preserve a memory of a measurement campaign through a physical 3D object. It is also an excellent teaching aid for explaining what we observe!

Let us take the spectroscopic binary Alf Dra as an example (alpha Draconis) which was the subject of a measurement campaign initiated by STAROS Projects. The spectra are freely available on the project website:

https://alphadra.staros-projects.org/database/

You can download the spectra of your choice to reconstruct the period graphically or in 3D. You can also use your own spectra, which will have been processed in the same way or retrieve from the BeSS database, series of spectra from a Be star, for example to highlight an outburst.

To produce the .stl file required for 3D printing, you will need two pieces of software:

• Visual Spec, spectral analysis software designed by Valérie Desnoux where she recently added a feature to generate a file containing the information needed to create a 3D view from a series of spectra:

http://www.astrosurf.com/vdesnoux/

• Open Scad, software that allows you to generate 3D files in .stl format. You can download it for free by following this link:

https://openscad.org/

Preparing the file with Visual Spec

• Define a working directory in Visual Spec in the menu **"Options"**, then "**Preferences**" and "**Working directory**".

In the same menu "**Preferences**", check in the tab "**Continuum**" the two normalization values that must be compatible with the series of spectra that you are going to submit to Visual Spec. The wavelength range of these two values must not contain any spectral lines, just the continuum of the "flat" star.

• Open all necessary spectral profiles. This may involve dozens or even hundreds of spectra.

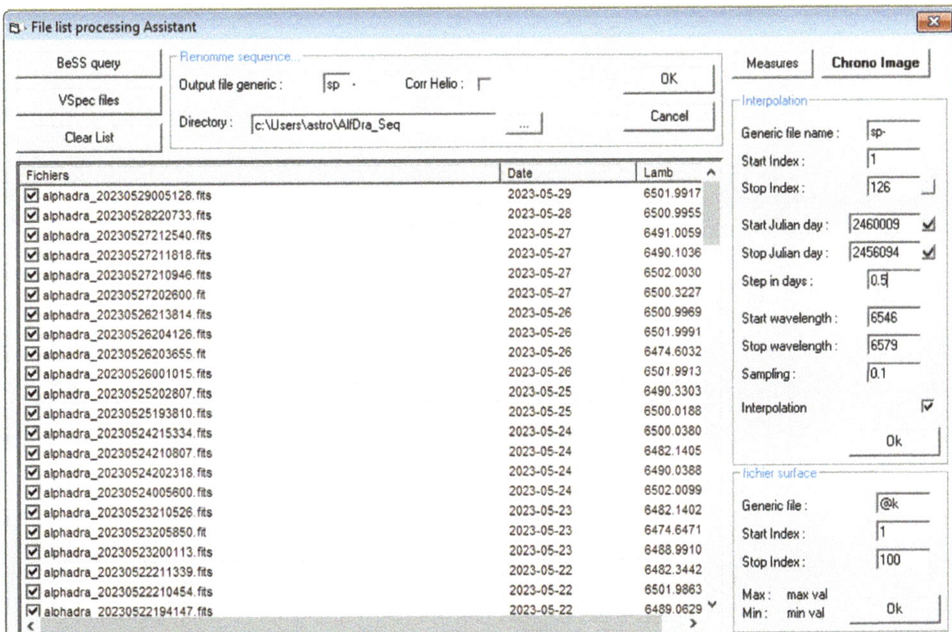

Fig. 10.11.1 — Window "File list processing Assistant". This window has three separate processing areas: the first at the top "Renomme sequence", then the one on the right "Interpolation" and finally at the bottom right "Fichier surface".

• Click on the menu "**Batch**", and then select "**File list processing...**".

• At the top right, you can import files opened in Visual Spec by clicking on "**VSpec files**". You can also request spectra from the BeSS database by clicking on "**BeSS query**".

• The window "**File list processing assistant**" opens. Indicate at the top, in the box "**Output file generic**", a simple name for Visual Spec to rename all your files in chronological order by date. In the example, sp was chosen. Verify that the current directory is correctly indicated below this box. Then click on the "**OK**" in the area "**Renomme sequence**". Visual Spec generates files sp-1.fit, sp-2.fit, ..., spN.fit in chronological order and writes them to the working directory.

• On the right side of the window, The total number of spectra and the start and end dates of observation are entered automatically. You just need to specify the desired step in days (0.3 in the case of Alpha Draconis) and the desired sampling of the wavelength representation (you can leave the default value of 0.1 Å).

The interpolation box can be checked: Visual Spec will then generate intermediate spectra according to the step size in day defined above.

• Click on "**OK**" from the area "**Interpolation**". After a calculation time that can be quite long, the software displays the dynamic 2D graph of the spectrum series.

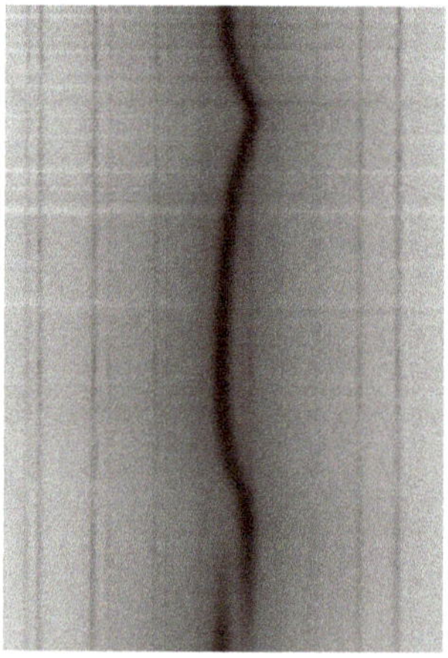

Fig. 10.11.2 — Dynamic 2D graph of Alpha Draconis. Generated by Visual Spec from 126 spectra in the STAROS database.

We can see that Alpha Draconis is far from being a sinusoidal curve because this spectroscopic binary has an eccentricity $e = 0.43$. A second companion can be identified by studying more periods.

• Finally, click on the "**OK**" at the bottom right of the area "**fichier surface**". Visual Spec will generate the surface.dat file that will be used by the Open Scad software to generate the 3D profile.

Printing with Open Scad

Then open the Open Scad software and enter this line of code:

```
scale([0.5,0.3,5]) surface(file = "surface.dat", center = true, convexity = 10);
```

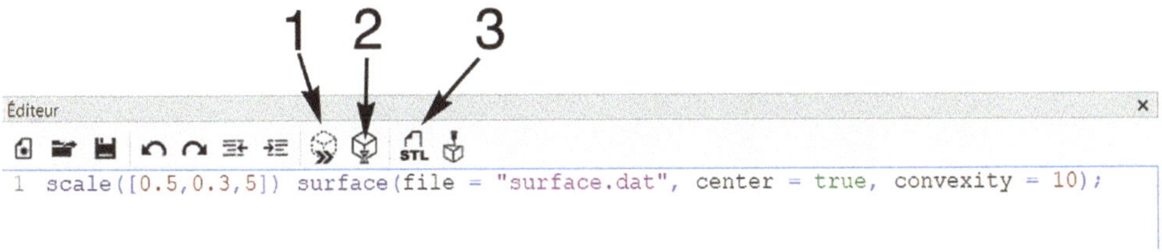

Fig. 10.11.3 — Open Scad software menu. Locate the three icons labeled 1, 2, and 3.

• Save the Open Scad program file in the current directory containing the surface.dat file.

• Click on the icon n° 1 as shown in Figure 10.11.3. The software generates a preview of the 3D shape of the alpha Draconis period.

Fig. 10.11.4 — 3D representation of the period. Generated with Open Scad software.

You can modify certain parameters such as the dimensions of the object to be printed by changing the X, Y, Z values of the scale function ($[X, Y, Z]$). You can also add to this object, a name, a logo as shown in Figure 10.11.5.

• Once you have finished making your changes, click on the icon n° 2 to generate the file (see Figure 10.11.3), then save your .stl file by clicking on the icon n° 3.

Fig. 10.11.5 — Result of printing the alpha Draconis period.

Your .stl file will then be retrieved by your 3D printer's slicer software to potentially manage layers of different colors and generate the G-code file. All you have to do now is print your curve in 3D.

10.12 Bibliography

- Astronomie & Astrophysique 4th edition by Agnès Acker (Dunod 2005) ISBN 2 19 995462 7

- SB9: The ninth catalogue of spectroscopic binary orbits by D. Pourbaix *and al.* (A&A 424, 727-732 (2004)) DOI: 10.1051/0004-6361:20041213

- Modeling and Analysis of Eclipsing Binary Stars by Andrej Prsa (IOP Publishing 2018) DOI 10.1088/978-0-7503-1287-5.

- Giants of Eclipse The ζ Aurigae Stars and Other Binary Systems by Thomas B. Ake and Elisabeth Grifin *and al.* (Springer 2025) ISBN 978-3-319-09197-6

- Ce que disent les étoiles by Danielle Briot and Noël Robichon (Belin Pour la Science Editions 2013). ISBN 978-2-7011-6285-0

- Spectral modelling of circular massive binary systems. Towards an understanding of the Struve-Sahade effect? by M. Palate and G. Rauw (A&A 537, A119 (2012)), DOI: 10.1051/0004-6361/201117520

- Visual Spec tutorial by Valérie Desnoux:
http://www.astrosurf.com/vdesnoux/versions/Vspec400.html

- Coordinated monitoring of the eccentric O-star binary Iota Orionis: optical spectroscopy and photometry by Sergey V. Marchenko *et al.* (Mon. Not. R. Astron. Soc. 317, 333±342 (2000)). DOI: 10.1046/j.1365-8711.2000.03542.x

- The frequency of spectroscopic binaries by Carlos Jaschek and Ana E. Gomez. (1969-1970 IOP Sciences)

- Sur la relation période-excentricité pour les binaires spectroscopiques by R. Bouigue, A. Pedoussaut and P. Prêtre (observatoire de Toulouse), 1967 IAU Symposium N° 30.

- Observation of exoplanets using the radial velocity method by Christian Buil:
http://www.astrosurf.com/buil/extrasolar/obs.htm

- Understanding the Lomb-Scargle Periodogram by Jacob T. VanderPlas (2018, The Astrophysical Journal). DOI 10.3847/1538-4365/aab766.

- The generalised Lomb-Scargle periodogram, A new formalism for the floating-mean and Keplerian periodograms by M. Zechmeister and M. Kurster. (A & A 496, 577-584 (2009)). DOI : 10.1051/0004-6361:200811296.

- The Scorpio-Centaurus association: I-Radial velocities of 120 bright stars by Wiliam Buscombe and Pamela M. Morris (1960MNRAS.121..263B)

- Introduction to the Study of Spectroscopic Double Stars by Guillaume Bertrand:
https://etoilesdoubles.org/revue/ED-2023-06/ED-2023-06-BERTRAND.pdf

- A Jupiter-mass companion to a solar-type star by Michel Mayor and Didier Queloz (1995 Nature, Vol. 378).

- The observation of exoplanet 51 Pegasi b by Christian Buil:
https://britastro.org/wp-content/uploads/2018/09/Observation-of-the-exoplanet-51-Pegasu b-V2.pdf

- Radial velocity measurements and orbit determination of eight single-lined spectroscopic binary systems by R. Bischoff (Astron. Nachr. - AN 1, No. 1, 1 – 6 (2017)). DOI: 10.1002/asna.201713365.

11. Measuring filters, slits and more...

This final chapter is devoted to measuring the bandwidths of astronomical filters, the width of slits, the wavelength of laser pointers or even the creation of liquid spectra, using a spectrograph that will be used "on the bench".

The concept behind Lab'Ex is to provide a compact, easy-to-use "laboratory" instrument with various modules available depending on what you want to measure.

To obtain the bandwidth of a filter, you can use any low-resolution spectrograph which allows you to observe simultaneously at once, the entire wavelength range of the visible domain spectrum (between 400 and 750 nm).

The autoguiding part of the "stellar" spectrograph is not used during this exercise and can be removed.

In "Lab'Ex" mode, the spectrograph will be mounted on a test bench. The method of use will be described for two common models used in astronomy:

1) With the Alpy 600: modified with a 10 μm slit, which increases its resolution to $R = 1100$ approximately. Used without its autoguiding module and without its calibration module.

2) With the Star'Ex: in low-resolution mode, equipped with a 300 lines/mm grating and a 10 to 14 μm slit to aim for a resolution similar to that of the modified Alpy 600.

The choice of these two instruments is not random, these are currently the most widely used spectrographs. They are both very compact and therefore easy to integrate into a test bench, unlike the much larger and bulkier LHIRES III or LISA.

For the spectral camera, if the exposure time is short (a few seconds), we can try using an uncooled camera with small pixels. However, conventional cooled CMOS or CCD cameras remain a safe bet to take accurate measurements without worrying about exposure time or ambient temperature.

Sensors with small pixels ranging from 2.4 to 3.76 μm like the SONY IMX 183, IMX 533 or IMX 585 are the most preferable.

In some cases, the use of higher spectral resolution may be considered, for example, for the examination of certain very narrow band filters, in order to improve the accuracy of the measurement of the effective width or centering on the wavelength cited by the manufacturer.

The use of the Star'Ex is highly recommended because you can easily change the grating. However, one of the advantages of low resolution is lost because the entire range of the visible spectrum will no longer be available, unless you use an Echelle-type spectrograph, which is much larger, more expensive, and requires an optical fiber.

Given these considerations, there is no need for a very wide spectral range to measure a small transmission width!

11.1 WARNING: Precautions to be taken!

In this chapter, we will use different light sources. Please note that some may pose serious risks to human and animal health or safety!

It is therefore essential to take precautions when using these light sources!

• UV lamps or LEDs emit light that is invisible to the human eye, but these wavelengths affect the retina and can damage it. It is therefore recommended to wear suitable glasses to block this radiation: UV-protective glasses.

• Laser pointers, regardless of their color, also pose a significant risk because all their power is concentrated on a single wavelength. They can cause permanent burns to the retina, or cause certain materials to combust.

• Wear special glasses adapted to the laser wavelength emitted (Be mindful of the light path and any possible reflections!). Note also that certain lasers such as some green lasers may also emit signifiant invisible but hazardous infra-red laser light.

• It is also advisable not to leave lasers switched on for long periods of time, their lifespan decreases rapidly with prolonged use!

• Some sources can also become extremely hot, Infrared or laser radiation may melt certain exposed parts of the Lab'Ex, given its plastic construction!

• The electrical voltages used for certain lamps (220 V) can also be dangerous, It is advisable to use a 12 V / 220 V converter where the output current is limited and to properly insulate electrical wires to avoid the risk of short circuits.

• It is also advisable to operate the Lab'Ex in a dark room in order to avoid light pollution: this could alter the results! (Pay particular attention to desk lamps or daylight!)

• Do not leave the sources unnecessarily illuminated. Illuminate them only for the duration of the measurement!

Assuming that the readers of this book are reasonable and thoughtful people, the risk of problems remains low provided that basic safety rules and common sense are observed!

However, the risks are real, so we urge you to remain very cautious when handling this equipment. The authors do not accept any liability for the consequences of any injury or damage!

11.2 Building your own filter test bench

Lab'Ex is a modular system comprising several parts:

• The bench: made from an aluminum profile to which the various modules and the spectrograph are attached.

• A light module: to produce the light source required for measuring a filter (light source free of spectral lines).

• A filter holder module: which can accommodate any type of filter up to dimensions of 50 x 50 mm.

• A test tube module: for capturing the spectrum of various liquids.

Users can supplement Lab'Ex by designing other modules depending on the type of measurement witch is intented. For example, a laser pointer mount, or a fiber optic module for analyzing the spectrum of objects or minerals.

All 3D printing files (.stl files) and assembly instructions for each module are available on the website:

http://o.garde.free.fr/astro/Spectro1/LabEx.html

11.3 The Lab'Ex with an Alpy 600 or a Star'Ex?

Depending on the spectrograph you have, you have the choice of making a mount for an Alpy 600 or for a Star'Ex.

Both solutions have their advantages and disadvantages.

Fig. 11.3.1 — The Lab'Ex with an Alpy 600. Here equipped with a cooled CMOS camera ZWO ASI 183 MM pro (pixels size of 2.4 μm).

The version equipped with an Alpy 600 is very compact because the spectrograph has a diameter of 50 mm and a length of 57 mm. The camera is usually larger than the spectrograph. The only drawback is that you can not change the grism (transmission diffraction grating + prism) to a higher resolution grism.

The Alpy 600 has a fixed resolution of approximately $R = 600$ if you use the Alpy 600 based on its original 23 μm slit (mirror slit supplied with the autoguiding module) or from 25 μm (round nickel slit included with the basic version of the Alpy alone).

It is possible to increase this resolution using a 10 μm slit (Shelyak reference: SE0253). In this case, you would need a camera with small pixels of 2.4 μm is recommended to ensure sufficient sampling.

If you already own an Alpy 600, this is undoubtedly the most economical solution for building a Lab'Ex.

Fig. 11.3.2 — The Lab'Ex with a Star'Ex. Here equipped with an uncooled ASI 178 MM CMOS camera (2.4 μm pixels).

If you do not yet have a spectrograph, the Star'Ex version will be the least expensive solution. It is easy to change the grating and thus have a higher or lower resolution.

It can be built without its autoguiding module and in the low-resolution version, that is to say with two 80 mm focal length lenses and a 10 μm slit or the 4-position slit at 14, 20, 26, 32 μm which is specially designed for this instrument.

11.4 The light module

The light module is the essential component of Lab'Ex. It must produce a homogeneous, diffuse light without spectral lines. LEDs or video projector lamps are not suitable. Halogen bulbs are preferable as their profile is close to that of a black body (Planck curve).

The disadvantage of standard halogen lamps, is that they heat up and, above all, that the available flux in the blue part of the spectrum is quite low. To improve this photon deficit, several dozen exposures will need to be taken with levels of around 50 000 ADU on each one.

There are two solutions for implementing this lighting module: either the version we designed, which is available via the general link to the Lab'Ex project mentioned above, or the version proposed by Christian Buil described on his website:

http://www.astrosurf.com/solex/specinti_annexe_fr.html

The module we designed has 4 xenon bulbs powered by 12 volts and a frosted glass to diffuse the light whilst allowing near-UV rays to pass through up to ± 350 nm.

Fig. 11.4.1 — The Lab'Ex light module. Equipped on top with a high-voltage power supply (250 V) for the calibration lamp (optional).

It also includes a wavelength calibration lamp which has sharp and know emission lines, powered by a built-in 12 V/250 V converter. Most small spectral bulbs can be inserted into it such as the Fireflies found on the conrad.com website or a classic argon/neon lamp but but it is not essential to add this component, which involves high voltage (and therefore is to be handled with care), if you already have a calibration lamp.

11.5 Measuring the bandwidth of a filter

The Lab'Ex filter holder module can accept filters of various sizes, including diameters of 31.75 mm, 2 inches, 36 mm unmounted, etc. or 50 x 50 mm squares. If you have other filter models, you can, of course, design a suitable holder that is compatible with the filter holder.

Fig. 11.5.1 — Example of a filter holder. Here is the version for unmounted filters with a diameter of 36 mm.

The filter holder module is installed on the Lab'Ex between the light module and the spectrograph. Measuring the bandwidth of a filter involves taking two series of spectra measurements:

1) The spectrum of light passing through the filter.

2) The spectrum of light alone, without the filter.

In both cases, The spectrum of a wavelength calibration emission lamp is measured immediately after the last exposure.

However, as we work here on the bench, the measurement system is very stable and it is not absolutely to recalibrate the spectrum for each measurement (the high-voltage part of the light module is optional).

Spectrographs are very compact and stable, they are not subject to variable mechanical stresses (as on a telescope) and the temperature of the room in which the Lab'Ex is installed probably does not vary much.

A "generic" calibration polynomial calculated at the beginning can be used and remains valid as long as the spectrograph or camera is not removed from the Lab'Ex.

However, if the grating is changed to one with a higher resolution, it may be necessary to recalibrate the spectrum for each measurement.

Since halogen lamps tend to heat up, this can also alter the spectral profile obtained. It is therefore desirable to perform the two series of spectra in as short a time as possible, one after the other.

In most cases, the exposure time will be quite short and will not exceed approximately 3 to 5 seconds given the intensity of the light source.

As the exposure time is fairly short and the spectral signal high, you can do without making darks frames, but you do need to take bias frames. The post-processing is carried out in the usual manner for the two spectra produced, except for the following points:

• Do not use a flat, as this is already present in the spectrum of the light alone.

• Since this is not a stellar spectrum, the two areas of sky background subtraction should not be done. By the way, that is anyway impossible, because the spectrum will occupy the entire length of the slit.

• Use a fairly wide spectrum binning range, typically 300 to 500 pixels.

• For greater precision, always take the centre of the spectral zone at the same value Y.

You can use any program you like to perform this processing but the Spec INTI software offers special functions for use in Lab'Ex mode (They are described by following the link above to Christian Buil's website.)

To obtain the response curve of your filter (its transmission curve), there is one last operation to perform: divide the two profiles using to this equation:

$$\text{Response curve} = \frac{spectrum_{filter}}{spectrum_{lamp\ alone}}$$

Spec INTI offers a function "**ratio**" which simply performs the division. With ISIS, you can divide the two profiles using the function **"Profiles arithmetic"** in the menu "**Tools**" then "**spectra 1**":

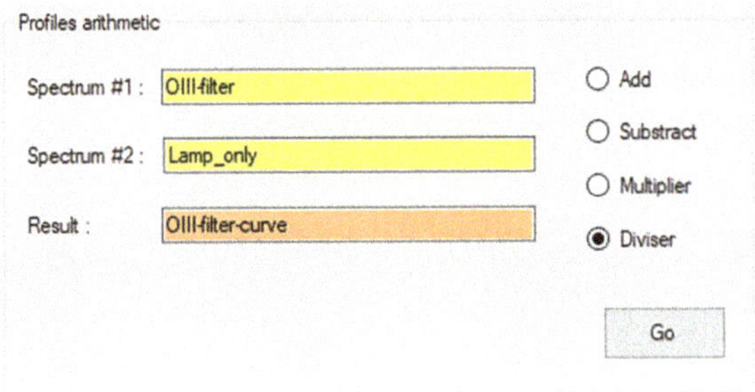

With:

Spectrum # 1: the name of the filter spectrum file, in .fit or .fits format.

Spectrum # 2: the file name of the spectrum of the lamp alone in .fit or .fits format.

Result: the name of the calculated transmission curve file: the result of the division, in .fit or .fits format.

11.6 Practical case with 3 narrowband SHO filters

There are a multitude of different filters available, including broadband, narrowband, UV cut, IR cut, order filters, and filters designed for visual use (UHC, "special planets") etc. All these can be measured with Lab'Ex.

Here is an example using three SHO narrowband filters (sulphur, hydrogen, and oxygen). These filters are very often used for imaging emission nebulae or planetary nebulae as the light of these objects is mainly concentrated on these three wavelengths. The SHO filters select:

- One of the two prohibited sulfur lines noted [S II] at 6 731 Å. It is the S filter.

- The hydrogen line $H\alpha$ at 6 563 Å. It is the H filter.

- One of the two forbidden lines of oxygen, denoted [O III] at 5 007 Å. It is the O filter.

In addition to the filter type, manufacturers also indicate its width (FWHM), which is often 3 nm, 5 nm, or 7 to 8 nm. This is also a parameter that can be verified from the spectra.

For each wavelength tested, a series of spectra is produced with the filter and then without it (just the lamp). Process the 3 spectra and divide each result by the spectrum of the halogen lamp alone.

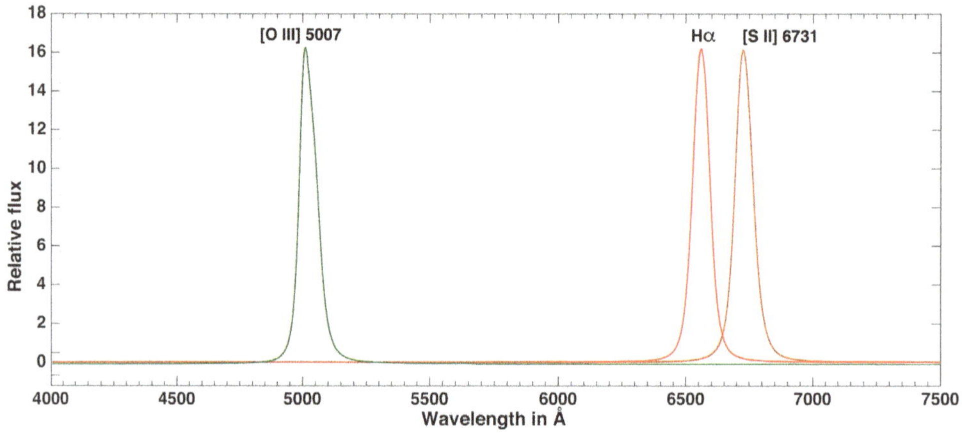

Fig. 11.6.1 — Transmission of the 3 SHO filters. The bandwidths specified by the manufacturer: [S II] = 8 nm, $H\alpha = 7$ nm, and [O III] = 8.5 nm (the 3 profiles are shown on the same graph). On the x-axis, the wavelength in ångströms. On y-axis, relative intensity relative to the continuum normalized to 1.

The representation of the three curves in Figure 11.6.1 allows us to verify whether the filter is properly centered on the nominal wavelength, which is the case here. High-resolution measurement would provide greater accuracy in determining the value of each transmission peak.

Measurement of the full width at half-maximun of each filter (FWHM of the transmission curve as we saw in section 8.4) allows you to obtain the bandwidth of the filter. It is common to encounter discrepancies between the manufacturer's specifications and the actual filter measurements.

In this case, the width values are significantly different from those stated by the manufacturer, namely:

[S II]: 8.11 nm (instead of 8 nm).

$H\alpha$: 7.6 nm (instead of 7 nm).

[O III]: 7.75 nm (instead of 8.5 nm).

Filters are often poorly centred on the published nominal wavelength.

The same procedure can be used on broadband photo filters (LRVB). Low resolution is preferable.

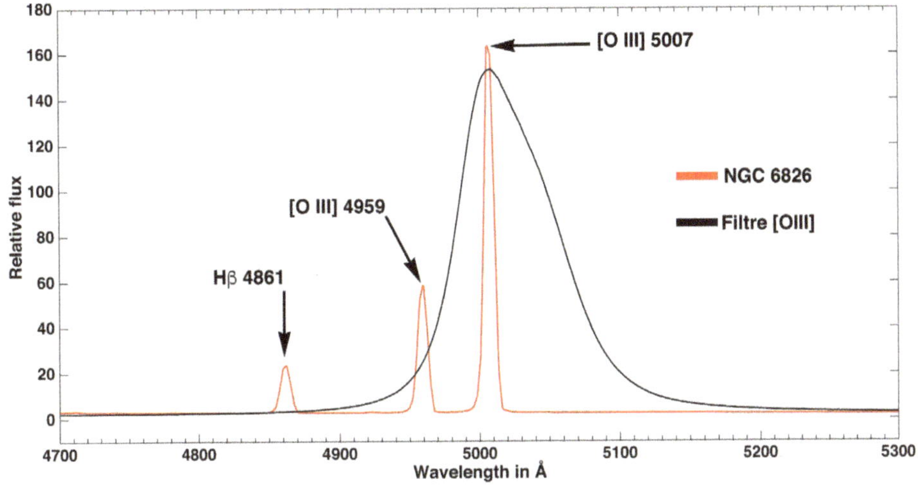

Fig. 11.6.2 — Details about the O III filter. In black, the filter transmission; in red, the spectrum of planetary nebula NGC 6826. Note: there is no physical relationship between the relative flux of the nebula and that of the filter transmission curve. On x-axis, the wavelength in ångströms. On y-axis, relative intensity relative to the continuum normalized to 1.

If you have in your archives the spectrum of a planetary nebula, it is interesting to superimpose it on the filter transmission curve. We can then see what the filter allows to pass through among the different emission lines of the nebula. A discrepancy may explain why some filters are more effective than others. This is often due to the very delicate manufacturing process: thin film deposition.

11.7 Measuring a light source

The Lab'Ex can be used to measure known light sources such as neon lamps, laser pointers, light bulbs, or any other sources whose nature is unknown.

Determine the correct exposure time, as some sources can be very intense. It is not uncommon to have exposure times in the order of a thousandth of a second. In this case, you can also reduce the gain of the CMOS cameras to make it easier to avoid saturating the image.

Depending on the nature of the source used, you will probably need to build a specific support (for example, for a laser pointer) in order to position it correctly in front of the spectrograph slit.

The source must be far enough away from the spectrograph so as to obtain a homogeneous light flux covering the slit or place a diffuser between the lamp and the spectrograph.

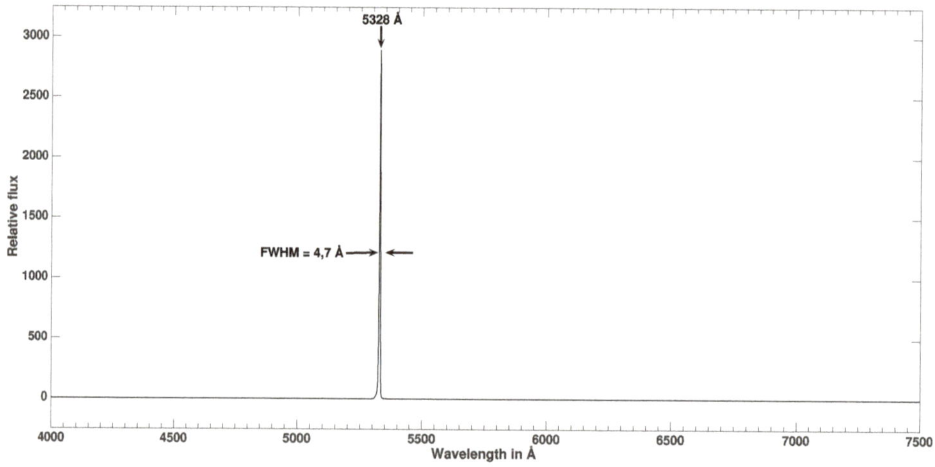

Fig. 11.7.1 — Transmission of a green laser pointer. These pointers have a nominal wavelength of 532 nm. Measured here at 5 328 Å, either 532.8 nm. On the x-axis, the wavelength in ångströms. On the y-axis, relative intensity relative to the continuum normalized to 1.

Measuring the wavelength of a laser pointer (LASER, an acronym with the words: "*Light Amplification by Stimulated Emission of Radiation*") can be very informative in terms of knowing its real value and not just its nominal value. It can be important for measuring the width of a slit (see section 11.9).

Most of the time, for inexpensive laser pointers, the wavelength tolerance is in the order of ±10 nm relative to the nominal value. It may also vary with operating temperature of the laser.

Other light sources can be measured and identified such as spectral neon, argon/neon, or other gas bulbs, LEDs, or any other light source whose spectral profile you wish to determine.

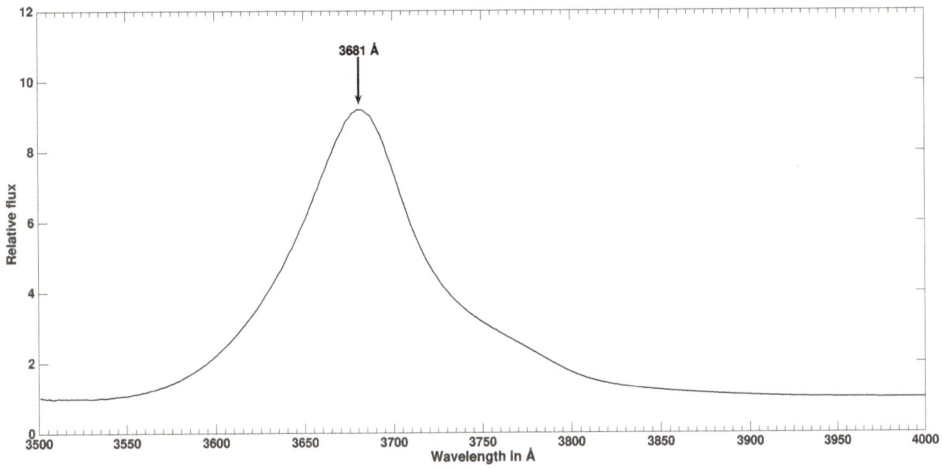

Fig. 11.7.2 — Emission of a UV lamp around 365 nm. This lamp is used to check the markers on banknotes: UV light makes printed fluorescent marks visible. Reminder: this invisible UV radiation is harmful! Remember to protect your eyes and do not point in the direction of people or animals. On the x-axis, the wavelength in ångströms. On y-axis, relative intensity relative to the continuum normalized to 1.

The near-UV lamps that are commercially available, consist of LEDs emitting in this wavelength range. In this example, we can see that the nominal wavelength of the lamp (given as 365 nm) has a peak emission centered on 3 681 Å, a 3.1 nm difference, and that the width of the emission band is sufficiently large, with an FWHM of more than 90 Å or 9 nm.

11.8 The spectral signature of a liquid

The Lab'Ex test tube module can be used to produce spectra of liquids in order to identify certain chemical elements or colors. To perform, wavelengths in the near UV or infrared range should be selected to identify molecular bands which are not as common in the visible range. This would allow us to discover whether wine has been chaptalized or adulterated!

The liquid module test tube transmits near-UV radiation up to 350 nm, It is more than sufficient for the use case we have chosen.

Let us take a well-known liquid as an example: olive oil, which we will analyze in both "white" light with the light lamp module consisting of 4 xenon lamps, but also with a 365 nm UV lamp whose wavelength we determined in the previous example.

Here we must digress briefly on the subject of fluorescence

Fluorescence occurs when a molecule receives light, absorbs it, and re-emits it at a different wavelength, ever longer and sometimes into the visible, thereby indicating the presence of this molecule. This technique is widely used in medicine, hydrogeology, fraud prevention, etc. It is also the "black light" used in nightclubs!

Fluorescence observation should be performed perpendicular to the excitation beam for greater accuracy (that way you can only see the light that is re-emitted).

The acquisition of spectra and their processing is identical to that described for filters. The only difference is that we will use two different light sources.

We will therefore produce four series of spectra:

1) A first series with the light module lamp and olive oil in the test tube.

2) A second set with the light module lamp and an empty test tube.

3) A third series with the UV lamp, which should be positioned perpendicular to the incident light as shown in Figure 11.8.1 (without removing the light module).

4) A fourth series with the UV lamp and an empty test tube.

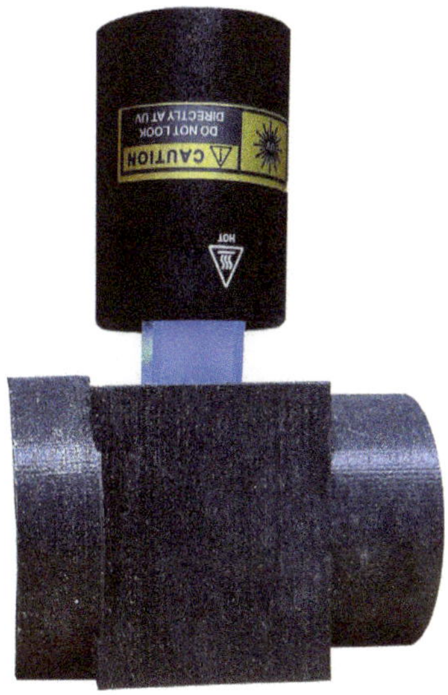

Fig. 11.8.1 — Using a UV lamp. Position the lamp on top of the test tube. (Remember to protect yourself from UV light.)

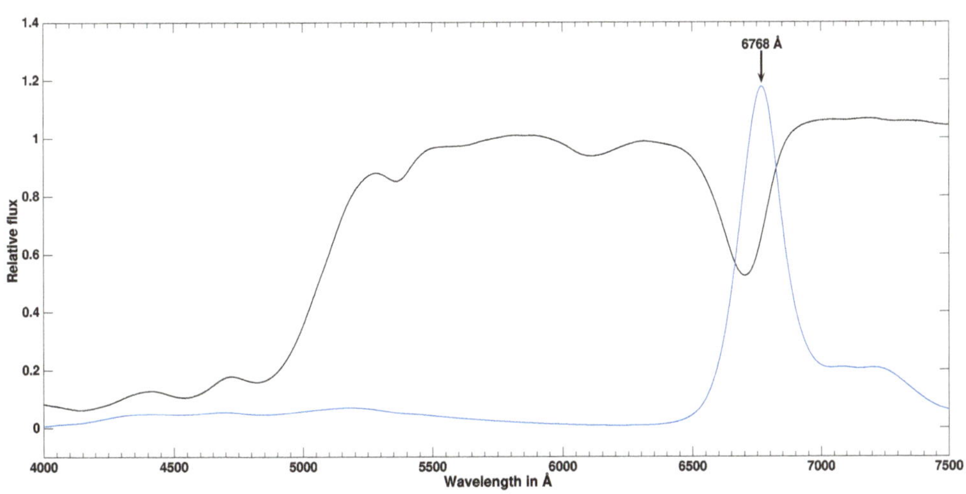

Fig. 11.8.2 — Olive oil spectrum. In the visible range with a halogen lamp (black curve) and a UV lamp (blue curve) that shows a peak in the red at 6 768 Å. On the x-axis, the wavelength in ångströms. On the y-axis, relative intensity relative to the continuum normalized to 1.

The transmission curve with white light shows a deficit in the red range and in the blue part of the spectrum, which gives olive oil its yellow color. With the UV lamp, fluorescence is highlighted together with an excess of red light is present, corresponding to the excitation of chlorophyll A and B molecules.

These molecules are rather green in visible light (see Figure 11.8.2). But when they are excited by a UV light source, they re-emit photons in the red part of the spectrum: the peak is here at 6 568 Å. As with stars, this "spectral signature" allows us to determine the composition or purity of products!

Other plants, meats, milk, and cheeses also contain molecules that are sensitive to excitation by near-UV radiation. Many books and publications provide valuable information on this fluorescence (see the bibliography section at the end of the chapter).

11.9 Measuring the width of a slit

This last measurement does not use Lab'Ex. It is included in this chapter because it remains within the field of spectroscopy and can be very useful, because it allows you to measure the width of a hand-made slit "home-made" or "DIY" but also to understand the diffraction of light.

Some slits do not have markings indicating their width. Others do not necessarily have the nominal size specified by the manufacturer. Sometimes you may be unsure about the width of the slit you have in your hands.

If you have a good microscope, it is possible to measure this width, but there is a less expensive method that can be applied simply with a laser pointer. It is a source of monochromatic light, usually of colors red (at 650 nm), green (at 532 nm) or blue (405 nm) with a tolerance of around ± 10 nm.

The label on the pointer usually indicates this wavelength. If you do not have this information, you can measure it with your Lab'Ex (see section 11.7).

The method of the measurement consists of illuminating the slit to be measured with a monochromatic light source (in this case, a laser), by projecting the diffraction pattern created by the slit onto a screen, which will vary depending on the width of the slit.

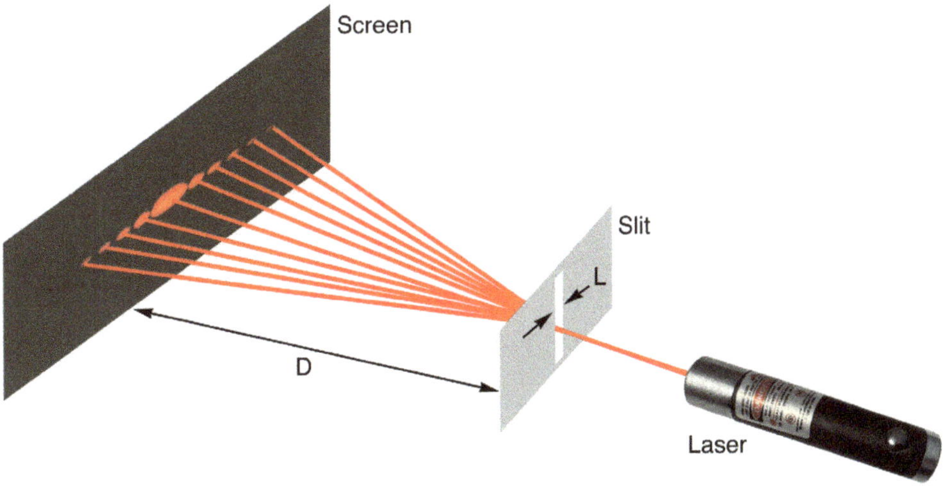

Fig. 11.9.1 — Principle of measuring the width of a slit. L is the width of the slit in μm and D the distance between the slit and the projection screen in meters.

The main problem in carrying out this measurement is having an ergonomic and rigid test bench. Laser pointers are often cylindrical in shape, They must be able to be rigidly fixed to a support that is integral with the slit to be measured.

A small assembly was made using 3D printed components which place the laser on the same support as the slit. The creation of a universal support is hardly feasible because the slit plates vary too much.

Some are circular with diameters between 10 and 40 mm, others are rectangular or square, measuring 15 to 30 mm. You will therefore need to adapt!

This test bench is placed at a distance D from the screen, generally between 1 and 2 meters away, in a dark room to see the light diffracted by the slit clearly. The thinner the slit is, the weaker the mark on the screen will be. The slit must be perfectly parallel to the screen.

The pointer push button will be locked in the "ON" position, with a piece of tape or a clothes peg! The distance in millimeters between two diffraction nodes is simply measured with a ruler.

To determine the width of the slit, the following equation is used:

$$L = D\frac{\lambda}{\Delta x}$$

With:
L: slit width in micrometers (μm).
D: the distance in meters (m) between the slit and the projection screen.
λ: the wavelength of the laser in nanometers (nm).
Δx: the distance in millimeters (mm) between two diffraction nodes.

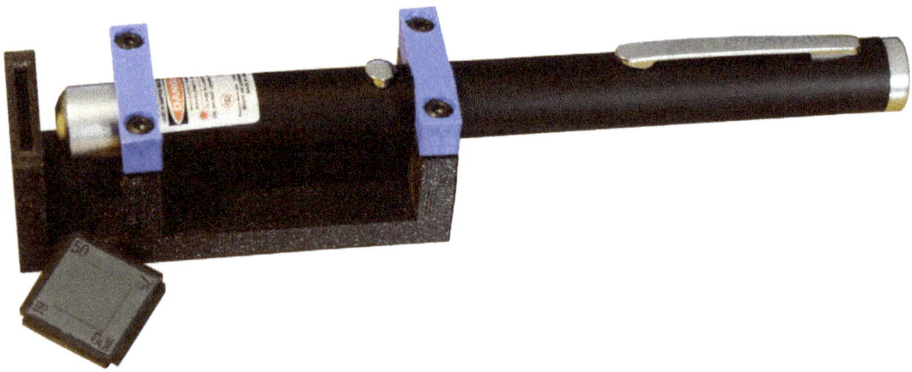

Fig. 11.9.2 — 3D printing test bench. Here, the bench was built to measure Shelyak Instruments' 4-position square slits for the LISA, LHIRES III, and Star'Ex spectrographs.

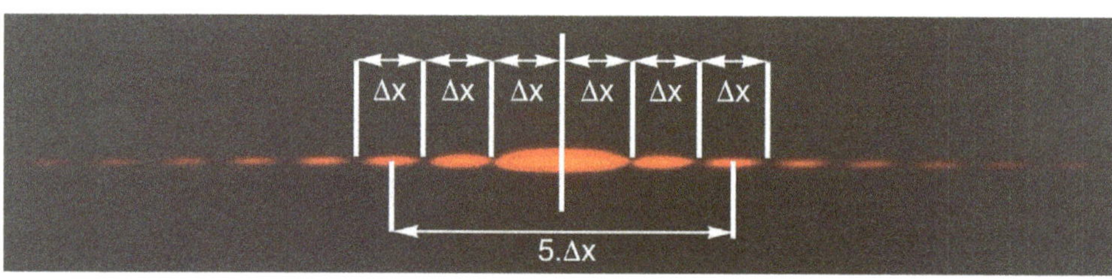

Fig. 11.9.3 — Diffraction pattern measurement. The value to be measured is denoted by Δx (in millimeters). For more precision, the measurement can be taken on all 5 strips and divided by 5 to obtain the average value of the length of a fringe. By definition, they are all the same size.

The width of the slit is found with a measurement uncertainty of approximately ± 0.2 μm (using the wavelength tolerance of the red laser at 650 nm (± 10 nm based on the model used here).

This error can be reduced by measuring the real wavelength of the laser used with a high-resolution spectrograph such as the LHIRES III, Star'Ex HR, the BACHES or the eShel (see section 11.7).

The table in Figure 11.9.4 shows the values Δx measured with a red laser at 650 nm for various Shelyak Instruments slits.

Slit width	$D = 1$ m	$D = 1.5$ m	$D = 2$ m
10 µm	65 mm	97.5 mm	130 mm
15 µm	43.3 mm	65 mm	86.7 mm
19 µm	34.2 mm	51.3 mm	68.4 mm
23 µm	28.3 mm	42,4 mm	56.5 mm
35 µm	18.6 mm	27.9 mm	37.1 mm
50 µm	13 mm	19.5 mm	26 mm
75 µm	8.7 mm	13 mm	17.3 mm
100 µm	6.5 mm	9.8 mm	13 mm

Fig. 11.9.4 — Table of measured values. Depending on the distance D and the width of the slits with a red laser at 650nm.

You can also have fun with lasers of other colors, The slit width which you determine should be the same!

11.10 Bibliography

- Lumière et luminescence 2sd edition by Bernard Valeur (Belin Science Editions 2016). ISBN 978-2-410-0071-9.

- La lumière et la vie, une subtile alchimie by Bernard Valeur and Elisabeth Bardez (Belin Science Editions 2015). ISBN 978-2-7011-7688-8.

- Optique, une approche expérimentale et pratique by Sylvain Houard (Boeck Editions 2011). ISBN 9-782-8041-6399-6.

- Fundamentals of molecular spectroscopy 4th edition by Colin N. Banwell and Elaine M. Mc Cash (The McGraw-Hill Companies 1994). ISBN 0-07-707976-0.

- Numerical Python in Astronomy and Astrophysics by Wolfram Schmidt and Marcel Völschow (Springer 2021). ISBN 978-3-030-70346-2.

- Spectroscopie atomique, instrumentation et structures atomiques by Émile Biémont (Boeck and Larcier editions, 2006). ISBN 978-2-8041-5035-8.

- Spectroscopie, cours et exercices by J. Michael Hollas (editions Dunod 2003). ISBN 978-2-10-007071-8.

- Evaluation of the overall quality of olive oil using fluorescence spectroscopy by Elena Guzman, Vincent Baeten, Juan Antonino Fernandez Pierna and José A. Garcia-Mesa (Food Chemistry Vol. 173, 2015, pages 927-934).

- Fluorescence of vegetable oils: Olive oils by Maurizio Zandomeneghi, Laura Carbonaro and Chiara Caffarata (PubMed 2005). DOI: 10.1021/jf048742p.

www.ingramcontent.com/pod-product-compliance
Ingram Content Group UK Ltd.
Pitfield, Milton Keynes, MK11 3LW, UK
UKHW061956041026
15591UKWH00016B/1024

9 782759 840090